THE FIELD

NATURE'S ETERNAL LINES OF FORCE, THE SUBSTRATUM OF ALL SCIENCE AND REALITY

By

Perry W. Peña

ISBN

9781365898259

Table of Contents

PREFACE

HOW DO YOU KNOW? WHAT MAKES YOU CREADIBLE ON THIS TOPIC?

My credibility starts with years' worth of in depth of study and reading of all the main philosophers and scientists here discussed. All the sources of deeply research work will be found in the bibliography but that is only half of the story of who is communicating this information and why you need to listen. It starts with a bit of a tragic true story which is facetiously entitled a THE TRAGEDY OF THE RETARDED GENIUS.

Let's join this story back in 1990, this is the year that I, Perry W. Peña (Pronounced *Penyu*) was at the end of my freshman year in college at the University of New Mexico. Funny story, and entirely embarrassing as I am the retarded-genius referred to in the title of this work. In that year at college I was failing and otherwise doing very poorly compared to my peers to the point that the difference was stark and obvious. I found myself needing to negotiate with all of my professors to let me make up assignments and tests and some acquiesced to it. This kindness helped me make at least c's, passing grades in most classes, but none of my scores were remarkable. I was mortified by the time May came around that first semester. I came to the solid conclusion that I had to have a real learning disability, a mental handicap, a limitation, a mild retardation or some other processing failure like dyslexia etc. It just had to be the case. I believed I was in need of a reader, or a software assist or several assigned personal tutors.

By May of 1990 I acted on my suspicions, I checked myself in to the UNM Assessment and Counseling Services center for a complete test of IQ and academic abilities and or limitations or handicaps so as to detect any mild retardation or learning disability. This is all a true story so stay with me. When I went to the diagnostic center I was subjected to the full battery of intelligence *and* academic ability testing and processing. The results knocked me on my ass! The final conclusions showed that my academic outcomes were the product of high intellect united with extremely poor education produced simply "average out" c level work at best. Academically I had high features and low features which produced a wash. That was the academic picture supplied by the testing. Superior intellect married with superior education would have yielded superior academic outcomes. This was what was uncovered but now the mystery of what caused this anomaly was just beginning to unfold. Now the tests that were administered which gave reality context to me were these...

- Wechsler Adult Intelligence Scale-Revised (WAIS-R)
- Woodcock-Johnson Psychoeducational Battery-Revised Pt. II: Tests of Achievement
- Nelson-Denny Reading Test
- Diagnostic Spelling Potential Test (DSPT)
- Woodcock-Johnson Psychoeducational Battery-Revised Pt. I: Cognitive Ability
- Bender Gestalt Test

Evaluation Completed 5-21-90 by Robert R. Romero M.A., Certified Educational Diagnostician

Mine is a cautionary tale. I am by definition, and compartmentalized category, a *genius*, according the UNM's diagnostics and aptitude testing. Of all the tests administered at that time only the *Wechler* and *Bender Gestalt* matter for understanding my intelligence apart from my education levels and for understanding the type of intelligence which I possess. These revealed a "superior" intellect which numerically correspond to an IQ between 125 and 149 according to averages on the accepted intelligence scales such as the Levine and Marks Scale for example. Howard Gardner identified eight types of intelligences and mine was determined to be *audio-visual* intelligence. Other aptitude tests performed determined that I am at the *"genius"* level with "spatial awareness". Now the net effect of my public school administered by lesser intellectual "educators" produced a gaslight effect which essentially ruined, undermined and cancelled, or short circuited, my first instincts. Gaslight education sewed self-doubt or diffidence into my intellectual instincts. I came to distrust my own perceptions and perspicacity and grew dependent on the judgements of others. These are people whom I would have to put trust in to best guide me through life, math, science, relationships, perceptions of God, good and evil, right and wrong and in the general guidance of the course of my life. I became gullible, vulnerable, a gudgeon a mark and easy target for manipulation. My strengths were essentially turned against me and my potential success. My strengths and insights were crushed down and driven inward to the only place they would be tolerated by the world.

My public school education was incredibly deficient! The public school system was simply not equipped to deal with someone of my category. No accommodations were made for me. In fact no attempt was made to understand my thinking. My public school education actually harmed me, it caused me to doubt my instincts and to fully adopt a fundamental *diffidence* of all my critical thinking instincts, intuitions, observations and insights. University testing revealed an intellect near 150 on the high side of the scale range. This I.Q. did not avail me nor propel me to auspicious heights. My self-doubt was so thoroughly inculcated into me that I came to perceive an opposite of reality regarding myself. One product of this "education" rendered me as nothing more than a sidekick to other lost misfit kids. My insights have always been those of an alpha so it has been incredibly humiliating to play the role of an ancillary. I did this with whoever was willing to hang out with me as isolation was just too painful.

So complete was the undermining of my sensitive psyche that I literally came to believe completely that I was the antithesis of an intelligent, credible, authoritative person or a person with any trustworthy intuitions with no valuable input! The destruction of a potential good life, an elevated life, was totally successful. My diffidence manifested reading skills which were low and slow possibly to a fourth grade level. The same goes for my math. For all twelve of my public school years and one in college not one single teacher or adult in my life tested my intelligence nor was anything done by anyone in my education to even bring me up to grade level at least. In fact several of my teachers diagnosed me to my parents as a spaced out "daydreamer" who doesn't listen! Not one noticed why I was exhibiting the learning problems I was and not one "expert" spotted the signs of something else which I was exhibiting! None of the experts, no "gifted teacher" intervened nor noticed that I was displaying audio-visual superiority, unusual spatial awareness and that this demanded an individualized education plan for me in order to ensure I could enjoy a successful educational experience.

My upbringing was a perfect storm of circumstances to complete the crushing of my superior intellect. The school successfully cast doubt into me, and into my family about me. It worked against me that some in my family were recognized as being of a higher than average intellect themselves, clearly above the 100 average and this worked against me and this is how. Because key people in my life were more intelligent than the average persons they also had it acknowledged by those of lesser intellect. These are the same people that could not comprehend an intellect which was a quantum leap above what they could understand as being already high. The peers and educators who gave recognition and respect to those around me as being smarter than themselves are the same who undermined my status as being intellectually superior. The educators who haphazardly dispensed labels and judgement on people's intellect became the arbiter of intellect to which we all adjusted our self-perceptions off of or based upon. While the educated and educators were correct about some of the family members around me, they were proven to be deeply wrong and flawed in their assessment or lack thereof concerning me.

No one noticed that I needed scads of encouragement and instead the majority of my teachers and mentors did the opposite. The majority of my insights were mocked in childhood and even on into adulthood. I now see that most of those who did this did so because they simply could not see as much as I could, nor could they hear as much nor feel as much nor perceive all the subtleties which I was picking up. I noticed this as all the people who were intelligent and who assumed that they were more intelligent than myself were the same people who could not comprehend how I could produce art that floored them and confused them on how such things could even be produced. I read of Leonardo Da Vinci and noticed that Da Vinci was misunderstood in the same exact way. I see (accurately) more than people intend to show, I hear (accurately) more than people say. I see more, I hear more, I perceive more and I have always been mischaracterized because of this as being "stupid, confused, misguided, mislead, simply wrong" or even "just crazy" in my perceptions. As a child I needed protection, advocates, I had none.

The battery of tests which I was subjected to at the University clearly revealed I was primarily an audio-visual learner. The tests categorized by audio-visual intelligence to be a *"superior"* mental processing abilities. This designation is tied to a number scale, there are more than one set of numeric designations for the classification of "superior". According to the varied scales a "superior" intellect is a numerical range of 125-149. 149 is a percentage so rare that is only found present only in 1.1 of the human population According to the Levine and Marks scale or 3.1 of the entire human race on the Stanford-Binet scale. No wonder that even normally intelligent people could not see it. Above average people are used to being the top of the scale and they see all deviance from their pattern of intellectual norms as belonging to those of others lower than themselves. I bought this assessment as well, I fully accepted it and integrated and internalized it as fact. Any intellect above normally accepted high intelligence proved recondite for others to perceive.

Think of what I was asking of average intellectual people, I expected them to understand, recognize and harmonize with someone with an intellect which is found, literally, not in a ratio of one in a million but rather an intellect found conservatively in only one in two billion two hundred fifty-eight million and sixty-four thousand and sixteen point thirteen. The number is even more isolating on the higher number ratio of the scale. Only 1.1% of the seven billion humans on earth can relate to someone with a 149 I.Q. Such a person with such an IQ is *literally* one in six billion, three hundred and sixty-three million, six hundred thirty-six thousand, three hundred and sixty-three point six four. These numbers make relating to other intellectuals difficult and expecting them to relate nearly impossible. Even for these much of what I reveal strains credulity for them. By extension the ratio scale between myself and common, 100 and lower IQ people, is a difference comparable to that of a human trying to have a meaningful relationship with a quadrumanus simian.

The puzzle which confused me as a kid now makes sense, and so do all my "teachers" and all the adults in my life. The fact that my visual processing was superior at the same time my auditory was superior meant that I had to tune one out and that most of the time during instruction, in most settings, I was literally shorting, or short circuiting my ability to absorb information. When I try to learn as others, sitting and listening my visual processing was in overdrive, needing instruction or other stimulation and it went to my imagination to supply it. My auditory faculties did similarly when the pacing of the auditory information was coming in slower than a pace that I needed to absorb the information at. My mind with these levels scared me as these two processes were over active and would generate endless visual scenarios and voices and full blown conversations in my head that made me sure that I was insane, I wasn't, I was misunderstood, my genius was considered an evil and bad thing by most and they led me to conclude the same thing of myself.

Until I was about 15 my competing audio-visual processes caused something like epileptic episodes that would last up to five minutes. I had many of them in class but I *never told anybody* what I was experiencing! My visual and auditory processes would go into a hyper mode and everything visually and auditory would rapidly expand and contract, move close and far, in and out! This is what was happening and no one ever helped me, no one created an I.E.P. for me. No teacher, counselor, diagnostician, child psychologist, principal, nor "reading expert" caught any of this all the years of my education. No one figured out that when I close my eyes I was absorbing complex lectures and prodigious quantities of information most of the time and most of the time I was punished for doing this. At other times I was able to absorb prodigious quantities of visual processing information when a process, such as in art was demonstrated in steps to me. I visually dismantle the world and put it back together in pieces visually and no one figured this out. I could imitate and form and simulate things seemingly straight out of thin air to most people's perceptions. They knew nothing of the process by which this happened.

After learning the reality of my life, I was driven to supplement all of my deficient education, which was not supplied to me the way I actually learn. From far behind the curve and far behind on life's starting line I self-educated and rapidly closed the gap on my education. Since then I have even leapt beyond the average college level of vocabulary, math, science, literature and reading and I intend to push even further. Since those discoveries I have put myself through college, through the University with a Bachelor's in Education, attending full time while working full time as a night janitor at Cleveland Middle School. I graduated having made the Dean's List for academic excellence. Further I have written and or published twenty nine books on historic subjects involving science, philosophy, Spanish colonization and the complex history of Hebrew colonization as well as the Spanish Inquisition. Other topics researched and written on involve the anatomy of martial arts, Biblical translation and to what degree the Biblical record is historically sound and accurate. Other subjects explored have to do with the Political parsing of history of the constitution and which political party has been more faithful to its tenants.

If this story describes you, if you can related Go GET TESTED!

INTRODUCTION

THE BIRTH OF TRUE SCIENCE & DEPARTURE FROM IT

The title of this book is "THE FIELD". Modern physicists study "fields" and "Field Theory" in relation to EMR or electromagnetic radiation. They often describe localized fields, around EMR or around electrons or electro-static phenomenon. This goal of this book is to introduce a lost and forgotten understanding of the Field, the grand Field which is the SUBSTRATUM of all reality, a field of fields and fields and energy grids within grids. This is the underpinning to all reality and science that was first explored by the following Greek thinkers. These great minds, with their logic deduced understandings and actually opened wide, explained and organized the mysteries of all of nature which we still wonder about today. Many believe that the big religions of the world and our governments are keeping incredible secrets from us, things we instinctively need to know.

Once you have read through this volume and followed all of the logic based revelations you will know as much as and more than all the religions and governments do. Governments are very much the same today as they were thousands of years ago. So be aware that the thoughts and logic derived conclusions in this volume contain the ancient scientific and philosophical wisdom of the ancient world. Alexander the Great, possibly the greatest conqueror and purveyor of government across the globe, was taught and personally tutored by Aristotle in most of the same content that you are about to learn. Much of this volume is an expansion from what Aristotle taught Alexander. With the animus of this particular wisdom Alexander went forth and became the world's most famous and powerful conqueror, becoming "Alexander *the Great*". The following is stated for the purpose of helping the reader properly adjust your assessment of the value of what is opened up for all to see in this great work. The Field may become the most important book in your life!

Before Aristotle was Plato and Socrates. Before their time there were the mind blowing philosophers known as the "Pre-Socratics". What the Pre-Socratics opened up came down to and through Plato and Aristotle and down to us in their writings. Let's begin our exploration of the underpinnings of true science, and philosophy, what they called "Natural Philosophy". Let's begin with Thales of Miletus.

Thales of Miletus 624-545 B.C. (E.) Circa.

(Pronounced *Thayleeez*), the first Scientist/Natural Philosopher who moved away from superstitious explanations of why things exist to a reasoning and logical empirical evidence based approach.

Greece was the birthplace of our modern scientific and rational thinking. The origins of the scientific method which modern science still uses begin there. Since ancient times Metaphysics and Physics and scientific empirical investigation were not partitioned things the way they are in the modern world. Any good Classical education began with a careful study of the philosophy of the Greeks. Such an education probed deeply the logic and rational derivations of understandings of Thales (pronounced *Thayleez*) of Miletus. Thales is truly the first philosopher and rational scientist of Greece nearly two thousand six-hundred years ago! Starting with Thales came reasoning and observation of nature which stated that human reasoning was all that one needed to understand the mysteries of nature and that no reference or interpretation of a god or superstition was needed. This was a first, a huge departure from the dominant thinking of the surrounding society of his time. The society of Thales' time ascribed EVERYTHING to the action of gods and spirits, even in the rocks and trees. When someone tripped over a rock in Thales time, people would grow suspicious of the rock over which they tripped as if the rock tripped them, plotted their hurt. So it was before Thales. After Thales science was born, math, logic, empirical observation and the rudiments of the scientific method were brought into the world. The mindset of Thales time would literally try to interpret why a rock would attack someone and then would proceed to plot revenge against said rock or attempt to appease the spirit of the rock.

Thales was different, Thales observed, reasoned, measured. Thales was one of seven of the ancient sages of Greece, each known for a wise saying or line of thought. Thales' was that "Water is best". The reason for this statement was based on careful observations of his own environment and careful note taking. Thales was inclined to state a hypothesis that water was the primary substance out of which all other forms of matter are derived from. Thales' reasons were very logically arrived at as Thales could observe water easily in a solid, liquid and vaporous state. Thales made observations and drew conclusions that life springs from moisture and definitely believed that he was observing spontaneous generation of molds etc. from moist containers. Until Louis Pasture proved otherwise, through modern scientific scrutiny and experimentation, in the eighteen sixties, all people believed as Thales taught. Thales worked on questions like, "Why is the world here?". And, "what is the world made of", and "by what processes?".

Thales put his logic and science to work, he used logic and math to figure out the height of the pyramids and to chart the stars for predicting solar eclipses and solstices and equinoxes. There are anecdotal stories of Thales using his wisdom of science and measurement of the stars to predict harvests and consequently know to prepare to exploit bumper crop situations for huge profit but only as a demonstration of the superiority of applying logic. Not bad for someone from nearly six hundred B.C. Thales was also a teacher, a founder of learning institutions and he had a protégé who was named Anaximander.

Anaximander of Miletus student of Thales c. 610-546 B.C. (E).

Anaximander was amazing and our first reasoning of evolution began with Anaximander. Anaximander stated, through observation, that more complex life comes from less complex life forms. Anaximander observed the fossil record in the rocks from long ago, sea shells on mountain tops and was able to observe some of the record accurately. Anaximander concluded that humans as well as all land creatures came from fish earlier in an evolutionary process which he concluded rose from the moist and muddy primordial sludge of the sea shores. Anaximander also observed correctly that our world, our earth, is just one of many worlds and that it to, the planet evolved as did all the others. This is not all Anaximander made many scientific observations and considered what we call the states of matter as elements intermixing to form the world we know. Even though Anaximander discerned the states of matter as "elements" he stated further that neither water nor fire were the foundational underpinning substance from which all things are made from. Anaximander, like all the Greek philosophers, took measurements and used math to aid them in observing the world. Doing as his teacher Anaximander took measurements that led him to understand that the earth is curved! This led Anaximander to arrive at the conclusion that the earth is rounded but he did not fully arrive at the conclusion that the earth is round or spherical, he speculated that the earth was round like a cylinder. Through logic and observation Anaximander used *inductive* and then *deductive* reasoning on his observations of the "elements" and could see that not any one could be the one from which all the rest were made from. Anaximander by extension then disagreed with Thales' hypothesis that water was the primary substance. In brief, Anaximander logically walked his reasoning to a conclusion that there exists a primary Aether substance underlying all of the states of matter or elements.

Anaximenes of Miletus, Era of 546 B.C. (E.)

Following Anaximander was his student, <u>Anaximenes</u> who was less distinguished than Anaximander and he represented a step back away from the conclusions of an underlying substance and reverted to Thales' observation only he differed in concluding that <u>air</u> is the primal substance which formed all the rest. He explained this by stating that fire must be "rarefied" air, water is thickened air and which still thickens further into land and lastly stone at its hardest. Anaximenes theory is very nearly the same logic path as that of Thales only concluding a different primary element as the foundation of the rest. Another important name occurring at this time was Heraclitus who also represented a step back to Thales in attempting to make one of the known elements into the foundational element. His element, was fire, Heraclitus believed that fire was the fundamental element because as he observed, all things change and nothing remains the same, like fire formations. Heraclitus was a bit like Empedocles in his belief that all of creation is a product of oppositional forces. This is an amazing observation. Heraclitus believed that war was the father of all, that it was a good thing, a perfect example, tension and of all things held in balance presenting as allegedly at rest like a strung bow, but actually still under great tension and opposition. This logic gave room and rise to another important departure from Anaximander's thread of investigation which caused us to question reality at another level.

Heraclitus 540-480 B.C. (E.)

This next philosopher was Parmenides, who can be seen as the antithesis of Heraclitus as he stated that the investigation of nature, for us, involves the use of our senses in order to even gather and analyze empirical data and it is a massive assumption to believe that our sensory experiences are right, accurate, trustworthy, real and true. Parmenides stated that we have no way to know if what we perceive as "real" is not all provided for us as an illusion or hallucination, a vision like that in the movie The Matrix or that we are not in the dream of someone else dreaming us! This leads to different conclusions, if this world of perception is imaginary then indeed our conclusions would in fact be delusions on our part and that what conclusions we arrive at are already pre-determined to end up where the dreamer was going to send us to all along. In other words, nothing is real, nothing really changes, nothing dies, nor is born, nor grows but is more or less a digital projection, a pre-made movie for us to watch and live in. Amazing, this line of reasoning is different from the pure observational science being developed by the others but incredibly important in this discussion and it to needs to be addressed as to where it fits in this line of investigation. You will find in this volume that there is a real answer to Parmenides' challenges. Parmenides and Heraclitus need to be studied together as both made very accurate observations of nature and reality that should not be ignored. One said nothing is real, nothing changes, and all is here at the same time because time isn't moving and that we are only imagining things to be in motion, birth death, change and what is was always going to be. Heraclitus told us just the opposite, he said everything is made of the transient element (fire) and that nothing is permanent and change is all there is. Who was right? This is where our brains should go to work as Plato's did, it is he who took up the so called "Paradox" and demonstrated how both are right in his theory of forms which we discussed briefly. Later, empirical science would show us that in fact Plato is correct. Our material universe is a product of a cause and effect reality. The material, matter, which breaks and builds on the scaffolds of the electromagnetic, manifesting the temporary which lasts a brief time, breaks down, all its parts are recombinant in nature and with be recycled and used over and over for ever changing new combinations. The electromagnetic does not degrade nor change, it is only impeded and it is the skeletal structure on to which atoms, molecules etc. form around to model something permanent and already present and a history already recorded in a steady state, a reality and stand-alone consciousness in the Electro-Magnetic Radiation Spectrum.

On return to the pure observational scientist/philosophers we circle back to Anaximander who was hottest on the trail in his observations of the need for an element apart from the known ones as a thing which underpins all the known ones. This is a powerful observation, a revolutionary leap forward in observation on the part of Anaximander.

ATOMISTS

Concurrent with Anaximander were two very scientific philosophers who through mathematic models concluded that yes indeed all the elements we know are constructed from something primordial. Their conclusions were that the underpinning element was the tiny indivisible shapes of the substances we see on a larger scale. The philosophers <u>Democritus</u> and <u>Leucippus</u> conceived of the theory of atoms. These two concluded that when you cut an apple or any other substance in half again and again that you will arrive at a solid indivisible, infinitesimally small foundation known as the atom which is a solid in contrast to the "void" which suspends the atoms. If all this sounds familiar it should, Albert Einstein and the academic main stream which gushed after his explanation of physics essentially revived atomism, the ancient theory of Democritus and Leucippus. While atoms are real, what they are actually like at the dynamic level is only speculated on even now. The belief in void, the nothing, the non, is a key component of the Atomist understanding of physics both anciently and among the modern atomists. At this point it needs to be clarified that the later philosophers Plato and Aristotle destroyed and disproved the void of the atomists which is important to be stated for the record here and now. Why? Because, the way that Plato and Aristotle, using logic, disproved the void and the ridiculous misunderstandings of reality which grow out from subscribing to atomism. The proofs of logic presented by Plato and Aristotle did not only overthrow the void of the Atomists of the past but the logic proofs overthrow the misguided physicists of today.

This is where the modern physicist has made a radical departure from reality and the science of investigating nature's mysteries as the atomists begin to define *imaginary non-entity* phenomenon as real. The Atomists with their insistence on void, empty space eventually led to their assigning attribution to the "Non". The atomists gave birth to a mind virus which leads to the denial, ignoring and avoiding of real forces and replacing of them with the non-real ideas such as "time" and "space" as things, or "principals" by Aristotle. Both "time" and "space" are not real things. Time and space do not have properties, nor are they measurable as Einstein mesmerizes the modern monkey to think so. Their physics is dead. Their physics doesn't help the human race do, build, understand nor advance anything as it is based on measuring ghosts and goblins and fairies so to speak. This blind alley pursuit happened before and was dismantled by Plato and Aristotle as a duty and a reset to reality. Too bad people don't know their history and philosophy and logic proofs because they just don't care. As a consequence of this attitude of stupidity perpetuation we have to relive and relearn the lessons already learned and wait in a dark age's limbo for the light to reach our so called modern "scientists". The Atomists of Socrates' time were hated so much that some thought their work should be burned down to the last volume. Now this can get all a bit confusing as I name all these different philosophers. Many of these philosophers existed before Socrates or did their work before his time but some of these men coexisted contemporaneously and had contact and knowledge of one another. Sharing a time frame with Democritus was <u>Empedocles</u>.

Democritus and Leucippus 460-370 B.C. (E.)

Empedocles, Unsung Hero of Scientific Investigation 490-430 B.C. (E.)

Empedocles is another philosopher who disproved the void of the Atomists and who hated them for their illogic in the matter of atoms in the void. Empedocles was like another Anaximander in that Empedocles also observed the fossil record and also observed that other worlds exist and that life as we know it progressed upward from simpler life elements. Empedocles' evolutionary march was awkward however; it was a product of pure random forces. This error (or incomplete analysis) of the randomness of evolutionary processes would later be called out and be cross examined with rigorous logic by Aristotle. Empedocles is far more important to science than is ever acknowledged. Empedocles is the one person who solidified the identity of the "elements" as four, earth, water, wind and fire. Empedocles demonstrated that these "elements" are truly unique and different and in opposition to one another and therefore concluded that none of these states of matter should be considered as the main underpinning substratum element. He derived through logic that there absolutely has to be an under pinning element, a primary fifth element, which is common to the four sensible elements, as their foundational source. The logic of Empedocles is independent of that of Anaximander and yet he is arriving at the same conclusions through logic and he won't be the last to get to the same inescapable conclusion as Anaximander. It is odd how the pre-Socratic philosophy unfolded because Empedocles is not a disciple of Anaximander yet his scientific investigation arrived at nearly the same path, which in turn advanced the science in the direction it was traveling toward.

Empedocles, like Anaximander, understood that the worlds evolved and knew there was an underlying substratum elemental component. Empedocles further understood and classified the known elements, (what we call the states of matter *retained to this day* in modern science). Empedocles took the science further by realizing that these "elements" were competitive over one another and therefore could not be the primordial force and source of their composition but that they had common foundational substance (Aristotle's "Substratum"). This is the logical place where

Anaximander arrived at but *Empedocles observed further still*, that, the primordial essence (Anaximander's Aether, Aristotle's *substratum*) had to itself have to have a *two part component, a dichotomy, a conjugate relationship to itself, a yin and yang type of essence to it. Components held in relative opposition to one another. This logical line of conclusions is amazing and grossly understated in importance*. Empedocles has been missed by almost all of modern science and how he moved abstruse understanding of the underlying reality ever closer. He is missed and his contribution undervalued by the so called intellectuals because of how he named the opposing elements. Empedocles RIGHTLY observed and concluded (i.e. in accordance with the order of the scientific method) that oppositional forces formed the universe. Empedocles named the fundamental oppositional engine of creation "Love" and "Strife". Empedocles concluded that these forces of opposition generated the matter of the universe by first gathering all of that is to one finite, infinitesimally small point of "Love" as love is a bonding and gathering force but with equal and opposite tendencies strife forcefully pushed out (exploded) out all the matter to scatter it as far as possible! Empedocles just uncovered the "Big Bang" two thousand years before Einstein and the modern physicists and it is actually more accurate than the model of the modern physicists' model as I will elaborate.

Empedocles' model is superior to the modern physicist's model of the big bang because it describes something that Stephen Hawking and Einstein failed to understand and explain namely what initiated the big bang, or "What was before the Big Bang?". The modern physicist goes on at length to admit that they do not have a clue about what caused the big bang to happen and why it isn't happening over and over right in front of us if it could happen once out of "nothing" then why doesn't it just happen again and again. They do not know, and both Einstein and Hawking throw up their hands at this point and state that perhaps this is the domain of God and they leave it at that. This is the death point of which I spoke of, the death of empirical observation and investigation and of learning. This is the superstitious default of humans and the abandonment of reasoning for understanding which Thales began. These modern physicists have actually destroyed what physicists were working on and driving towards but this work will go where they failed to by returning to the thread of investigation of lost modern scientists who pick up the trail of reality where Empedocles left off.

Foolish scientists missed the science of Empedocles and his highest of value observations that the primordial or "substratum" of reality's composition is a fabric which is conjugate to itself. All the value of what Empedocles was dismissed by the dismissible "scientists" merely because he used poetic terms instead of mechanistic or material dispassionate nomenclature! Everyone lost the trail of the true scientific substratum, including Plato and Aristotle because, frankly, Empedocles committed the unforgivable crime of science; he was ahead of his time!

Plato (427-347 B.C. (E.) on the left with his star pupil Aristotle (384-322 B.C. (E.) in Athens

To be fair to Aristotle it is important at this point to clarify another misunderstanding about just how clear these ancient philosophers' observations of nature actually were. Modern scientists often scoff at the scientific accuracy of the ancient philosophers due to their designating the four known elements of earth, wind, water and fire as "elements". If people would bother to read the actual writings of Aristotle he explains this all very clearly that the so called four elements were *not* single elements to the Greeks but rather descriptions of the "solid, liquid, gas and plasma" states of all materials! You know, like modern physics does now.

In scientific investigation there is no higher crime than to run a hundred, two hundred, or a thousand or more years ahead of everyone else's abilities to see. This will get you dismissed by fools as a "fool", madman, one who simply misunderstands or is confused or out of your element. This is now and always has been the way of humanity, all its forward movement is also counted and forced to regression over in over in a cycle of evolution and de-evolution. This was ultimately the big crime of Socrates, the teacher of Plato, who did not add new theories to natural philosophy but rather questioned and closely scrutinized and exposed weak theories which angered many.

Socrates of Athens 469-399 B.C. (E.)

Empedocles was right, what he called "Love" and "Strife" are in fact the identified two components of what caused the big bang, a collapse of the Electro-Magnetic field which was static in our perceptive region. Imbalance in the field set in motion a prodigious collapse of implosion and funnel like collapse inward to a massive compression and elastic tension like never before the counter force sprang back and banged out the product of tangled oppositional lines out into the outer most portions of the field. What Empedocles called Love and Strife are identifiable as what we now call <u>Magnetism</u> in juxtaposed oppositional tension with <u>Di electricity</u>! The only other ancient philosopher to come close to

Empedocles' idea of juxtaposed forces that form the elements and their behavior was Heraclitus who believed similarly to Empedocles regarding opposing tensions.

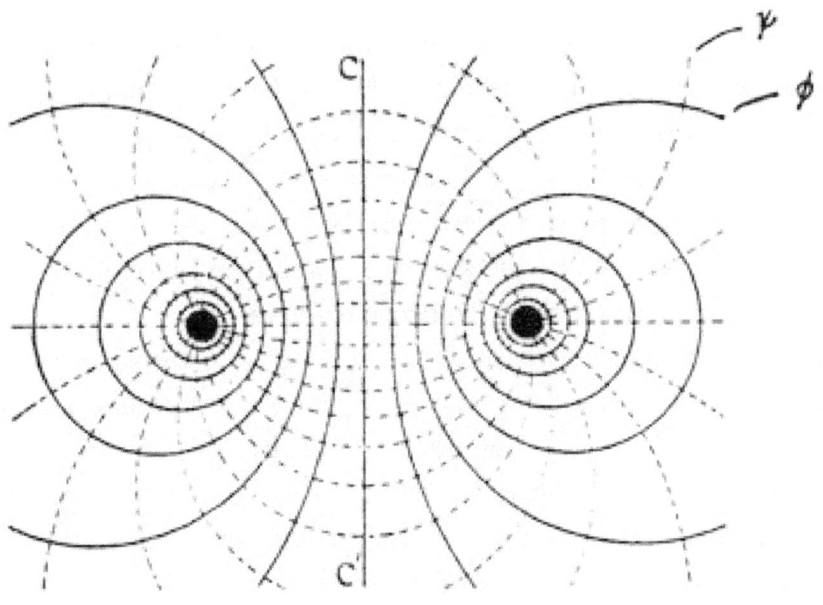

Modern field theory in electrical action, the conjugate lines of force which power our power lines look just like this, and Empedocles, an ancient Greek philosopher arrived at these elemental forces over two thousand years ago.

Modern scientists going back to Einstein knew of the reality of the electrodynamic forces which are Electro-Magnetic. Einstein and the British physicists were not working in, nor were they masters of the electrodynamic sciences. Frankly Einstein was not smart enough to work with his contemporaries who were Tesla, Steinmetz who were working off of the equations and observations of Oliver Heaviside, James Clerk Maxwell, Benjamin Franklin, Faraday, Galvani and Volta. This is sad, how more accurate could

Einstein's special and general theory of relativity have been had he worked with instead of in opposition and contradiction to the real working and producing scientists who are possessed of the arcane understanding of the real Aether. Instead of going to his betters, the real experts who could distinguish the real Aether from the vague erroneous understandings of it as the great endless field, Einstein chose to throw science back into the dark ages of mysticism by sighting the Michaelson and Morley experiment which he proclaimed to overthrow the existence of the Aether. Largely due to Einstein the Aether became **Bête Noire**. After this science was stunted and it all became abstract measurements of the *non-real elements* such as the curve of non-entities such as time and space! What time is was clarified and removed from the realm of mysticism two thousand years ago by Aristotle in his work the Physics and yet Einstein drug us back to the pre-Aristotelian mystic beliefs of time. Sad! Another mystic revival found in the special theory of relativity is the myth of the void, "empty" space as a void of nothing which he in turn reifies as a quantifiable entity with measurable and calculable properties such as to say space, nothing, has curvature and that alone explains gravity. Logic smashes the modern physics of atomism as it did the atomism of the past because both got into the business of reifying posterior attributes like those who consider a shadow an entity. Space is not a thing, a shadow is not a thing, a shadow is a privation of light. Curved space is like curved shadows or curved ghost aliens and magic unicorns. By this insanity Einstein unwittingly took us backward in science to the time of Sir Isaac Newton and his self-confessed lack of understanding of gravity. In the modern ignorant world, Einstein is held up and his name is made synonymous with genius when in truth, Einstein was no Einstein! I'm sorry, he did much more harm to empirical science than he did good and this is how he did it.

Don't misunderstand; Einstein did some very important gathering work and observational and theoretical work. One of the erroneous versions of the understanding of the Aether did need to be disproven and driven from the world of science. The problem was perhaps in the name Aether. By Einstein's time there existed multiple theories, descriptions and explanations of the Aether and most needed to be overthrown. By overthrowing one understanding of the Aether by this name he inadvertently overthrew all understandings of the Aether. This was not possible in reality as Einstein largely renamed, reused and repackaged the juxtaposed nature of the Aether with its electrodynamic equations. Electromagnetic field equations were ascribed as the equations which calculate black holes, gravity, and time and space warping and everyone thinks he is brilliant. It is a brilliant diversion and salesman job.

Perhaps it was for the best dispensed with the majority of Aether theories but not at the expense of dispensing with the EMR Cartesian Field version. The ancient scientists were exactly right, underpinning matter in all its states, underpinning atoms, is the field, the one boundless field which is simply a tangible fact. Even Einstein and Hawkins and all the modern physicists know there are magnetic and electrical fields, but this is where they lose the trail of reality, this is where they begin speaking of fields as <u>localized phenomenon</u> (*only present locus in quo*) and fail to recognize the grand field which is incommensurate even though it has been observed in optical holography. They simply do not know what they are looking at, they are looking down the barrel of the real Aether, a.k.a. THE Field! The property of the field was discerned logically by Empedocles correctly. Empedocles is the unsung hero, the unsung genius, his contribution has been relegated by the simple to the realm of minor historic contributions. Big mistake! The field is real, detectable, visible at points, boundless, elastic, conjugate (often misrepresented as + positive and −negative).

Aristotle brings to light some of the short sightedness of the observations of Empedocles in the areas of evolution. Aristotle points out and them proves that while evolution of life is evident and apparent it is not the product of randomness but neither does Aristotle then instantly defer any further investigation as to why life is here. Physics and biology are far more interlinked than either discipline currently realizes, as will be demonstrated in chapter 6. Aristotle points out that Empedocles missed the mark when he stated that evolution's products, the now extant animal species, were derived at by random forces doing the assemblage. Darwin later would tend to lean as Empedocles does on the subject but not committing to it with the same ardor.

Darwin knew what he was observing, he did not know then a thing about DNA and what it can yield and why it is doing that but he was hot on its trail with the best science available at the time. Darwin knew and cataloged change and change through mutation triggered by environmental factors. He was brilliant for his time. Aristotle was not mocking Empedocles but rather enhancing his observations and steering them more precisely to a better model of reality. What is missed by Aristotle and by Plato is how Empedocles did the same for their theories of the universe's evolutionary origin models. Everyone missed it, *well almost everyone...* If you know your Aristotle and Plato works then you know that both were convinced of the universe having an origin and a point and that the resultant formation of the universe manifest in a spherical expanse! Brilliant, correct, Einstein even agrees in the special and general theory of relativity. Where Plato and Aristotle miss the mark, where they default to human superstition is where they say this is where you find God, the prime mover, the cause, the first cause. Both agreed that the universe originated as a spherical expansion. Plato and Aristotle like many others fell into the trap that captures most, that tricky point of what was before the universe, before the big bang? They needed to look back to Empedocles for that answer.

When someone defaults to a god or gods or aliens or a prime mover, blind watch maker etc. one stops investigating and also negates all credibility because such logic never ever can answer why a God would suddenly and arbitrarily generate a universe out of nothing after endless eternity before 13.8 billion years ago. For such beliefs to come from men of logic who previously drove the point against the atomists that no void exists, that "nothing" is not a thing that exists and only nothing comes from nothing. Nothing never conceives and brings forth something. Parmenides in particular made the most powerful version of this argument and Plato and Aristotle were in full agreement with this point when it was aimed at the illogic of the atomists Leucippus and Democritus. This is where Aristotle and Plato missed where Empedocles made the proper scientific enquiry and deeper probe into the pre-origins of the material world with his primordial forces which he named love and strife, which is found in the place and conjugate

41

relationship where modern scientists find Di-electricity and Magnetism functioning just so.

Parmenides 515-450 B.C.

(E.)

Pythagoras 570-495 B.C.(E.)

This line of reasoning leads to yet another pre-Socratic philosopher and master of mathematics and logic, he is Pythagoras of Pythagorean Theorem Fame. Plato was heavily influenced by the Pythagoreans especially in the composition and explanation of his Theory of Forms. The Pythagoreans were masters of a secret, the ratio of life itself as discovered in the golden ratio and the golden section of the Pythagorean triangle. The Pythagoreans came to learn of a ratio which repeats in all facets of nature and natural proportions and behavioral growth patterns etc. It is amazing and not at all clear how they came to discover this ratio of nature. This ratio 1.618034, often referred to as *phi* factors in to the origins of the universe as the lines of force, the di-electric and magnetic come in the descending ratio which manifests as a phi to 1 descent in line pair sizes. The line pairs have always been tumbling as motion is an eternal given, when the big bang happened our discernable portion of the field was effected and caught up in turn in the tumble which triggered a magnetic suction vortex or compression and when the di-electric snapped back into position causing a massive blow out or push pull out expansion of the previously compressed lines of force. More on this in coming chapters. In a unbalanced field of differing descending ratios of phi to 1 motion is a given. The field has no final smallest diminutive descended ratio bottom. The travelling wave of motion from the impossibly small, literally never began and will never end.

Aristotle represented the final or ending of the great early deep thinkers and their community but this did not mean that their works were buried and went extinct after him. No Greek society and colonization and Hellenizing of the Mediterranean became the back bone of Roman logic and classical education. Until the fall of the Western Roman Empire, these great early scientists and philosophers were deeply integrated into Rome's education and thinking. With the fall of the Empire in the west, and the beginning of what has been called the dark ages, the great works of science and philosophy were lost and forgotten entirely until the time of the great western civilization's Renaissance which followed the crusades! This history is complex and the crusades played their role in the Renaissance. The Eastern part of the Roman Empire did not fall when the Western Roman Empire did, no instead it limped on in a highly contracted form well into the Renaissance and was in fact the source, the repository of the great thinkers in their libraries which fell to the Islamic uprising. Thus the great works of the Greek scientists/philosophers, as captured by the Muslim armies DIRECTLY led to the Golden age of Islam which spurred on an explosion of art, science and architecture in the Islamic world and this golden age predated the Renaissance in Western Europe. It is the crusades and the clash of cultures which generated the seeds of renaissance (Rebirth of learning) in the west. The reawakening of western civilization is tied to the great Middle Age thinkers, those who worked hard against superstition to restore civilization, learning, art, science, architecture, literacy, mathematics etc. The middle men rejoined the lines of study and thinking of the great Greek thinkers and then moved to advance from their last points of departure. The names of the middle men were names such as Rene Descartes, Copernicus, Kepler, Galileo Galilei, Leonardo Da-Vinci, Michelangelo, and of course Sir Isaac Newton. These men who follow represent those who would work the world back to accomplishments and observations up to Aristotle and begin the rigors of proofs through empirical observations. This truly does set our next group rightly in the middle of the understanding of the SUBSTRATUM but none of these men were yet able to take us up to the doorway that Empedocles once opened (The FIELD of opposing tensions). Not until we reach the moderns and their work with what they called "the Aether" do we properly understand the so called Aether to be a generative FIELD of Electro-(Dielectric & Magnet tension). The work of the modern scientists took two distinct paths, the biological (As Represented by Charles Darwin and Louis Pasture, Watson and Crick etc.) and the Mechanistic as opened up by (Franklin, Volta, Faraday, Galvani, Tesla, Maxwell, Heaviside, Morse, Thompson, Crooks, Farnsworth, Steinmetz. Hitherto the true link between these seemingly unrelated lines of inquiry will be merged properly for the reader's edification of how all the universe works together. These were the real Natural Philosophers and scientist, who were rooted in classical education who would carry the sciences forward with a real output, a real product (or products) which we all use and enjoy today and could not live without. Their work and line of pursuit all but died with them except for a few modern researchers and truth seekers who continued to further their research.

CHAPTER 1

On to the Observers and their Observations and also the Lack of Observations

The Middle Thinkers Gallery, Those Who Returned Us to Reason Who Began the Observations Leading to us Back to the Edge of the Substratum

Leonardo Da Vinci 1452-1519   Michelangelo 1495-1564

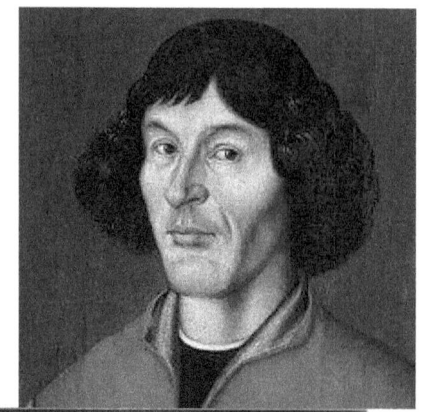

Nicholas Copernicus 1473   Johannes Kepler 1571-1630
Galileo Galilei 1564-1642

Rene Descartes March 31, 1596-1650

Sir Isaac Newton January 4, 1643-1727

Beginning with Leonardo Da Vinci the rebirth of science finally came back with force, Da Vinci was the most prolific creator, inventor and artist of the renaissance. Da Vinci has been described by artists as one of the greatest of all time. Scientists and engineers like to claim ownership of Da Vinci as one of the most impressive scientist/inventors/engineers of the renaissance period. Many biographers like to state that Da Vinci was a mathematician and scientist first and foremost and that he was only an artist second in priority. I love to read biographers who only know what they do about important historic figures from reading other misinformed biographers. They got Da Vinci one hundred percent wrong who say he did not value art first and foremost. To know Da Vinci and understand Da Vinci, you have to read Da Vinci in his own words. Yes Da Vinci wrote, and he wrote his thoughts on all subjects of his fascination, in his work THOUGHTS ON ART AND LIFE.

When we read Da Vinci in his own words he makes it clear that he was deeply studious and self-disciplined in math, engineering, science and painting but he did not partition these disciplines as his biographers did. Da Vinci repeatedly wrote at length of the *science of painting* and that to do it right one has to study and know the rigors of math, anatomy and proportions in order to do painting up to ones fullest potential. Da Vinci further stated that painting is a far superior form of communication to the human mind and spirit than is anything poetic or simple written or spoken words. Da Vinci is to be deeply respected even by the modern man as his rigors of study surpass the knowledge of the majority of even modern scientists. Da Vinci was brilliant, besides painting the world's most famous painting of all time, the Mona Lisa; he also invented flying machines, early designs of helicopters, gliders and parachutes. Also Da Vinci invented two robots that worked and were programmed to move independent of an operator. One was in the form of a suit of armor capable of sitting and moving its arms around and the other was a self-propelled cart which was pre-programmed to move on a course, turning corners etc. without a driver. Da Vinci invented a concept of a type of Gatling guns for rapid reload and firing as well as a tank and a giant crossbow.

Following the teaching of Aristotle closely Da Vinci understood and stated the scientific need of an Aether.

Nicholas Copernicus

While Da Vinci was a young man developing his ideas along came Nicholas Copernicus. Copernicus was another brilliant scientist who committed the unforgivable crime, Copernicus saw further ahead than his contemporaries, than the academic leaders of his time. Without the aid of a telescope Copernicus made careful eyewitness observations and mathematical calculations and came to the conclusion that the earth is not the center of the universe but that it rotates around the sun! Of course the scientists of his time said he is wrong, all of his peers rejected his work. The Roman Catholic Church denounced him for making such statements and did not allow his work to be read for almost three hundred years! So Copernicus was about three hundred years ahead of his time. So, as is true to form of the cutting edge and most gifted scientific minds, Copernicus also was clear about the Aether being a factor, but no he did not define it, nor understand it even as well as Empedocles. The story of Copernicus is both happy and tragic. Happy for us, he elevated us all out of ignorance of the solar system but this happened at the price of his being disrespected, shunned, dishonored and mocked and denied. Copernicus was alone!

Johannes Kepler

Kepler was the next genius to recognize that the earth revolved around the sun. Kepler was also resisted by his contemporaries and the church for this truth. Though Kepler, like Copernicus contradicted Aristotle's model of the solar system, he too could not conclude anything against the Aether.

Galileo Galilei

Galileo Galilei was another of similar conclusions but like Da Vinci was a tremendous artist and saw no line of distinction between art, math and science. This is if you believe the man's own words on the subject. Yes he too wrote a tremendous book called THE DIALOGUE. Galilei is most famous for his invention of the telescope with which he confirmed Copernicus' conclusions that the earth revolves around the sun and not the other way around. Galileo was threatened by the church as he was called before the holy office of the inquisition for implying that the earth is not the center of the universe! Galileo suffered for the truth, for being ahead of his time and being far more enlightened than his contemporaries. This genius also understood that there exists and Aether though he did not understand it well.

René Descartes

Here again is another great mind of the Renaissance; this was a philosopher and geometrician, scientist. He too understood the reality of the Aether, and never questioned it. The science which would demonstrate its omnipresence and conjugate nature was still far in the future from the time of Descartes. Descartes, like Aristotle knew to separate and designate the Aether/Sub-Stratum into the realm of Metaphysics as metaphysics is not tied and dependent on physics nor is derived from the physical universe. The Physical universe is the effect and the Field is the Cause. The physical reality is but a small bubble in an endless ocean/field. Descartes is most famous for his philosophical statement named "Cogito" which is usually translated as "I think, therefore I am". René Descartes gave us the "Cartesian Coordinate plane which models the grid of reality. The name "Cartesian" is named for Descartes. Descartes greatest works in math in science were completely predicated on the fabric of the Aether as our x, y, z coordinate plane! Frankly the Aether/Field is the inspiration for the

Coordinate plane employed by ALL mathematicians TODAY! Michael Faraday envisioned the Aether as a fibrous structure!

Newton

Sir Isaac Newton is yet another scientist who brought us closer to understanding the true nature of the so called Aether. The Aether was somewhat easy to overthrow as a scientific certainty by the time of Albert Einstein because it was so greatly undefined that it lent its name to many unscientific understandings of what its nature is like. The details of Newton's understanding of the field, or lack thereof are the subject of chapter 5.

Exit the Renaissance Enter the Modern Sciences, Enter Benjamin Franklin

Benjamin Franklin, yes that Benjamin Franklin of American Revolutionary fame was also a tremendous scientist and a very special scientist. Ben Franklin was a leader in the early science of electricity. It is Benjamin franklin who penned the longest standing rule in physics today that matter is neither created nor destroyed but merely transformed. Franklin made the key observations of electrical forces which helped us understand it, control it, contain it and use it. Electricity is very different than what scientists think today. Franklin made the key observations that the ratio of "electrical fire" moved from one location appears in a secondary location in the exact same ratio which it is diminished from the primary location. We have Franklin alone to thank for the positive and negative notation of electricity. Franklin has many inventions to his credit, one which would later be reinvented and modified almost two hundred years later in the form of the Van De Graph generator. He also invented the lighting rod and the electrical storm detecting meteorological device and many other amazing things which you were never told of! Franklin along with others taught us that electrical lines of force are ambient from the ground on earth and when pulled to tension when released return to ground. Franklin's work pretty much died with him and electrical advancements really didn't go too much further for another one hundred years.

Benjamin Franklin's work in electricity dealt in the realm of static electricity or dielectrics which is a field of study which has almost been forgotten due to the focus on EM radiation in all of its forms and applications. With the loss of study of dielectrics we undercut our ability to fully understand electricity.

Galvani and Volta

Luigi Galvani and Volta are two later scientists after Franklin who conducted new electrical experiments and it is Volta who essentially invented the power cell battery. Galvani, began to detect the real Aether, the Field, Galvani and Volta were beginning to understand that certain metals had an "atmosphere" around them, fields of static electricity or di-electricity was bending around the metals. Galvani started to experiment with the use of magnets adding a new component to the understanding of static electricity.

Samuel Morse and Oliver Heaviside

Morse's name is one you may recognize as he is the man after whom the famous "Morse Code" is named for. Morse was neither a scientist nor mathematician. Morse was a painter and an idea inventor. Morse came up with the concept of the Telegraph line and he paid for its construction and experimentation. He was in the time of Michael Faraday and James Clerk Maxwell the geniuses who did the math and calculations, on the Aether forces on THE FIELD which made the telegraph system work. The telegraph system did not work well at first; it had many problems, especially of degrading and weakening signals over long distances. When the telegraph was worked on to the point that it was successful on land it was proposed that a cable be dropped on the floor of the Atlantic ocean to facilitate telegraphing across the ocean. The electric, the static electric signal was degrading; it was failing to make it across the ocean. The problem was that at this time electricity was only half understood. The field turned out to be magnetic as well as dielectric and this problem was solved by the Aether manipulating calculations of the genius Oliver Heaviside. He too was ahead of his time and largely rejected and hated and forced to live a solitary life. These field Equations and manipulation of the Field proved what Empedocles said two thousand years ago, the field is conjugate and opposed in tension to itself and as a result generates motion and energy.

Charles Darwin

About the time of the telegraph's emergence another genius emerges, reviving and proving with empirical documented observations the notions of Anaximander and Empedocles concerning the origins of life. Darwin observed and dug up the fossils which demonstrated evolutionary change in animals branching off of what he called "the tree of life". This means that all living things, though appearing and manifesting different all come from a smaller pool of simpler life forms or a single form such as single celled life, up to fish on up to us. Science was making a tremendous come back in such a way as had not been seen since the renaissance! The picture of all things was beginning to come into focus and was now in a new infancy based on hard data and observation added to the older notions that proved accurate.

Louis Pasture 1822-1895

The work and research of Louis Pasture at this time, beginning about 1860, cannot be overlooked in importance. It is he who finally broke through in living and biological sciences with his research into vaccinations and bacteria control through heating. Pasture disproved an Aristotelian misunderstanding, he proved the law of Biogenesis that life comes only from other living things. Pasture and Darwin represent the living sciences and the bringing of scientific, empirical study back to biology. They broke through the superstitious explanations of the origins of life to take us back to Thales, Anaximander, Empedocles and then beyond them into the modern exploration of the DNA and RNA strands and code. Biology and the Electrical sciences are proving to have more in common than ever before realized. The story of biological origins and the E.M. Spectrum are closely linked and we can know all about the true origins of life and DNA when we learn about the FIELD, when we know what "material", or real substance(s) banged in a big way at the beginning of time. When we know what banged, then we know what states of things were formed. When we know how those two distinct formations act upon one another we will find our origins of DNA and subsequent "life", consciousness/intelligence imbedded in that rudimentary "life". Careful following this book and logic however or you too may end up committing the ultimate science crime! You may find that your scientific thinking is too far ahead of its time and that this information will not be accepted as everyday obvious truth for a thousand years.

Crooks, Tesla, Steinmetz and Thompson

These named scientists were brilliant and they carried electrical experimentation into the twenty first century. These, their work and experimentation caused and invented this entire electrical modern world in which you live now. Your, TV, radio, microwave ovens and microwave broadcasts (Like Wi-Fi), x-rays and alternating current trace back to these men and their successful understanding, calculation and manipulation of the great FIELD. They were on their way to giving us all free, limitless, clean energy once and for all... That is until all of their work and understanding stopped being taught all together as wrong, overthrown by Einstein and his followers. There are many books on all of these men, I highly recommend that you read them but always prioritize reading their own works in their own words.

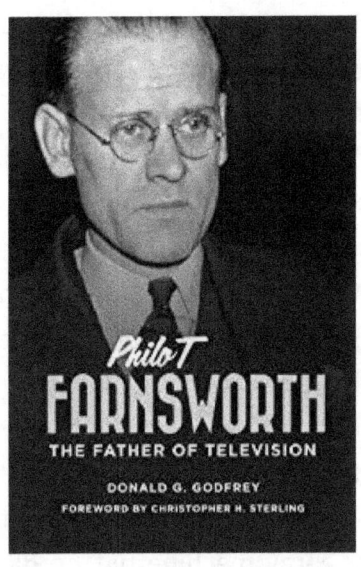

Philo Farnsworth August 9, 1906-March 11, 1971

Any mention of Crooks, and his vacuum tubes should include a mention of Philo Farnsworth, the man, whose research led to the creation of the first television broadcasts using vacuum tubes. The TV changed human civilization for all time! It is the most powerful media tool ever invented and the most powerful education/propaganda tool ever invented though the general manifestations of these tools have been tremendously abused and misused to the point of having the opposite of an educational and enlightening effect.

Tesla and the E.M. Radiation

Of all the modern scientists the least respected is the one who gave us the most and understood the Aether intimately as this is what his whole work was with and about. All that he was doing was experimentation on the Aether. Tesla was misunderstood to be completely focused on electricity. This is not so, his experimentation was on the full Electro-magnetic spectrum. Tesla constructed giant induction coils and he did this to experiment with the all the range of the E.M. spectrum. Tesla truly invented the x-ray machine, the Alternating Current motor, radio and wireless microwave transmission first as he had to sue to prove his claims. While Marconi is credited with the invention of radio and Wilhelm Conrad Roentgen for the x-ray machine these claims are historic lies and maladroit teachings and if you check the Supreme court rulings on these matters you will see. Not that Tesla pressed the issue concerning the x-ray machine but he did concerning radio. Tesla was getting all kinds of results in his labs which reporters observed and didn't know what they were looking at. Just as modern physicists still can't explain how Tesla was doing some of the strange things that he demonstrated in his lab for the elite of his day. I can. The explanation is the same one that proves that he was the first to discover x-rays and radio because they are all products of high and low coil frequencies! Tesla constructed large tesla coils which he exposed to fast moving magnets creating a rotating magnetic field trapped and guided by a static electric medium, the copper of the giant coil. Tesla would let current amplify in his coils until he shorted the coil flow at high, medium and low points. Tesla amplified the product of high frequency compounding further by running the coils together through Crooks' vacuum tubes, THIS PRODUCES X-Rays and even faster, more high energy, compressed E.M. Radiation expressions. This was direct experimentation and active braiding of the lines of force from large radio undulation to sub-atomic sized waves that may descend down infinitely. Because we do not detect, nor can we, the smaller wave forms, these can easily be expected to move much faster than the so called "speed of light". Tesla spoke of experiments with cosmic rays which moved many times the "speed of light".

Tesla's experimentations were not simply of "light" or of Electromagnetic radiation, they started this way but Tesla went on to understand that electricity, light and all forms of radiation were not subject to the limits which Einstein would tell us of. Tesla began to experiment with a form of electricity which was not limited by the speed of light. Tesla discovered, used and built devices, like his original radio design, which transmitted information by means of magneto-dielectricity vs. by means of E.M. transmission. These were Tesla's "Cosmic rays". Prior to Einstein's Special and General

Theory of Relativity Tesla's conclusion were verified by Professor Wheatstone, the uncle of Oliver Heaviside. Professor Wheatstone of Electrical Engineering fame, for the Wheatstone Bridge, calculated that dielectricity (static electricity) transmits at a speed which equals Pi over 2, times C (the speed of light is 186,000 mps) for a speed of 210,000 mps! This type of wave travel is longitudinal not transverse. This is *Magnito-Dielectricity*, not EMR. EMR is greatly misunderstood; it is misunderstood as "light" in all of its forms and size manifestations. There is no such thing as "light" as light is merely illumination, a product of EMR resistance. In space there is no light to be seen from the sun or the stars! Light is only seen as the higher, faster transmissions of longitudinal waves is slowed by collision with planets and are reflected to slow down to become visible light.

Speed of Light Nonsense

To be clear, there is no such thing as the speed of light. Visible "Light", makes up only a tiny fraction of the fuller spectrum of E.M. Radiation. If we call the full spectrum of seen and unseen radiation "light", then realize that radio waves are the long, tall and "light" wave forms, cosmic rays are much smaller and much, much faster and higher in energy if we are to believe Tesla. To talk of the "speed of light" causes confusion of the nature of the full E.M. Spectrum of radiation. For Einstein or any other physicist to talk to the public and the academic world of the speed of light is highly irresponsible and damaging to the investigations of just how fast super-micro lines of force can move before they shift to a section of the Field of much tinier dimension than we are capable of detecting. In the electrical sciences we see such a phenomenon occur and yet we hear nothing of this in the field of Physics. They ignore entirely! Electrical voltages can grow so powerful that they simply disappear for brief cycles in a "histolysis" phase where they depart from our dimension and go to a "place" which Electrical Engineer Eric Dollard likes to call "counter-space".

Of course the physicists try to explain what happens to energy at the point of hysteresis but their explanations utterly fail to explain the phenomenon. "Counter-space" is nothing more and nothing less than a temporary transfer or shift to micro lines of force which are much higher energy expressions of E.M. radiation fields than our dimension can handle. Energy is trying to match and find equilibrium and high voltage and explosions of various kinds do just that by shifting out during hysteresis just temporarily. For a moment, high force energy simply goes away, it isn't here where we are in a certain perceivable section of the Field. We perceive and live between a section of the great FIELD between what the most "brilliant" scientists consider Macro and Micro. Our ability to perceive the whole field is not possible! The boundaries which our so called scientists fix as the maximum of macro and micro compare to a single speck of dust measuring itself from one end to the other as it falls into the Pacific Ocean of endless dimensional downsizing and upsizing! Even this is a modest model used only to being the thinking of the immensity of the eternity which we are only a infinitesimally tiny place where our version of reality dwells! Yet this fits the Parmenides and Heraclitus paradox as we are both tiny to the point of being nothing and not existing and yet being everything of all importance. We are both, wrap your finite "scientific" little minds around this. No wonder the physicists are so afraid of the FIELD, it is maddening unless you embrace it as a whole!

Gallery of the Modern Scientists, the Good and the Misguided

Benjamin Franklin 1706-1790   Samuel Morse 1791-1872
Oliver Heaviside 1850-1925

Alessandro Volta 1745-1827   Luigi Galvani 1737-1798
Michael Faraday 1791-1867

Charles Darwin 1809-1882   James Clerk Maxwell 1831-1879
Nicola Tesla 1856-1943

William Crooks   1865-1923   J.J. Thompson

Charles Prodius Steinmetz 1832-1919   1856-1940

Marie Curie 1867-1934

Another very important research scientist to be noted is Marie Curie. Marie's research demonstrated that atoms are not the permanent things which previous researchers led us to believe but that elements have forms that are unstable and that can in fact decay and break down into simpler, more stable elements. Marie like many other scientists who went before her was ignored and harassed ripped off of her discoveries, which in her case was largely due to the fact that she was a woman.

According to the best of my investigations where science took its bad turn dates primarily to the era of World War II. This is not simply my opinion, this was the farewell message and warning of President Dwight D. Eisenhower as he was about to leave the presidency. Eisenhower warned that science, as a result of World War II the majority of the scientific community went into the business of obtaining government contracts to build and test new and better equipment and means to blow up the world! Eisenhower spoke of the pull of the Military Industrial Complex to become the growing and self-perpetuating force which is pulling all research and science in its direction and the commercial consumer manifestations side of the same among the public to buy sensational and vein products. This is where we are now as of 2019.

What happened in World War II to harm scientific thinking and investigation only culminated in and with World War II. The secret nature of the war technologies is what caused the split, the break with real powerful science and that which is taught to the public. The world famous Physicist Albert Einstein became the face of modern Physics and science in general when he released his Theory of Special Relativity which was very popular with the public and non-productive segment of the scientific community. The birth of purely theoretical physics came into it's own at this time. This is the time that endless long equations that describe properties of "space" and "time" were sensationalized and the quest for quantifying the entire universe began to take hold like never before! Behind the scenes when Einstein was growing in popularity the research was going on, the real work at General Electric Laboratories and later at the Naval shipyard at Bolinas California. What was happening in these places? This is where science was growing in leaps and bounds, much of this information was highly secret at the time but all can be verified today. These places are where the electrical sciences were being probed, documented and manipulated in ways you have never heard of.

As you may or may not be aware, one man, Nichola Tesla, invented the Alternating Current Generator Motor as well as many highly advanced electrical devices such as the wireless distribution of electricity and cell phone communication and the essence of the internet in heavy use in the modern world. These creations were conceived and built by Nichola Tesla as far back in time as the 1880s! You probably had no idea. This is only the beginning of this history and this long strange tale, which is not a fairy tale at all but the reality of this history. Nichola Tesla's contributions were amazing but there was a problem with all these things. No one else could make working versions in replication of Tesla's inventions and he himself lacked full ability, or will, to articulate how to exactly reconstruct his works for full scale industrial use and implementation.

This is where history takes an odd twist into a less well documented era in Tesla's life. Many people are very aware that after Tesla's Wardenclyffe tower was destroyed that Tesla was no longer in the public eye as he was before. This period has led many to believe that Tesla was all but done with his inventing. This is grossly inaccurate. Tesla continued to patent amazing, still not publically disclosed nor understood inventions. The misunderstanding of Tesla comes from people as a whole not being able to understand that Tesla's Wardenclyffe experiment did not fail but rather it was a success! The military became very aware of

Tesla's research at Wardenclyffe and the results of his work. The implications became very plain, what Tesla unleashed from the earth and how it was done proved something extremely dangerous. Tesla, completely unintentionally invented the ultimate doomsday weapon!

Tesla was not attempting to create a devastating weapon but non -the-less that is what he did and on the way to the success to this experiment many unintended revelations about the nature of nature and electricity's part in its composition came to light. In process of building up to the grand experiment Tesla himself was electrocuted quite severely by a stray jumping bolt which struck him, at high voltage, in the shoulder causing him incredible pain and something else. It is implied and inferred that when Tesla was struck by the bolt that his conscious mind was displaced momentarily and he stood or witnessed several points of time in his life simultaneously implying that the action of the bolt displaced him partially in time! This event is the basis for the theory that Tesla was involved in the Philadelphia Experiment. No actual clear documentation exists which clearly admits that such an experiment ever even happened and who in the scientific community were actually the involved in it. This is only speculative! The Philadelphia experiment was, as the story goes, was an experiment conducted in Philadelphia during World War II, on a navy ship named the Eldridge. As the story goes, the U.S. Navy was conducting and experiment to make this ship totally invisible to radar using extremely high voltage Tesla coils powered by large generators fed to a series of magnetic cables which lined the ship. As the story goes the experiment worked as intended but then something went very wrong. The ship it is alleged phased out of time and location and literally disappeared and this was caused by the extremely powerful electrical field. World War II stories like this persist involving the stolen research and plans of Nichola Tesla which are in actual fact MISSING!

Retracing my footsteps of how I started down this trail/rabbit hole of Science and Philosophy now seems to have no clear *locus in quo*. Perhaps as far back as the early 1990s I became interested in and focused on the late super human generous inventor, Nichola Tesla. It is at this time that I found and began reading the biographical work on Nichola Tesla, *Man Out of Time*, by Margaret Cheney. I had to read this book more than once because so much of what I was learning of in this work was largely unbelievable as I read about the man and his accomplishments and inventions. Much of what I was reading there underscored just how little of it that I understood.

Those Who Stopped the Scientific March and Reverted to the Physics of the Atomists

Albert Einstein 1879-1955   Stephen Hawking 1942-2018

In sharp contrast to the real research scientists who were mocked, rejected and abused for their whole carriers, both Einstein and Hawking were readily accepted, celebrated, lauded and praised as the most brilliant minds of all time by the likes of those with the least brilliant minds. What high praise! Modern versions of Democritus and Leucippus. Einstein's real genius is how he got rid of the eather field a dichotomy and replaced it with a dichotomy called Space/Time. Notice he retains the dichotomous nature of space while dispensing with the previously known dichotomy by largely doing nothing more than rebranding/renaming things already understood. By doing this he positioned himself as the authority of everyone else's previous work and understanding. His model of course is not just an innocent rebranding as it fails to account for matter's existence which separates physics into illogical compartments. Good job. He is a con man which fooled all those arrogant elitists who fancy themselves intelligent.

Very contrary to what you have been taught in classical physics, as descended through the work of Albert Einstein down to the particle physicists of today. These operate in a theoretical haze disconnected from the tangible production science of the early twentieth century. The universe and reality are built in an unseen field of infinite non-terminating, *unbounded*, lines of force whose composition consists of magnetic and dielectric (static electric) force in an opposed or juxtaposed orientation to one another. The field is a fabric, just like a woven fabric it consists of alternating magnetic and dielectric lines running in three dimensional configurations as illustrated mathematically as an x, y, z axis a.k.a. Flemings "Right Hand Rule". Sectioned off, or viewed or pictured in area isolation the field is essentially cubical. The field was not created and has no beginning and no end, it has no quantity and its existence as such contradicts the entire theme of Einstein based physics which is goal oriented in "quantify" the universe. This is why you hear the words "Quanta" or "Quantum" in the field of physics today. The field has NO QUANTITY, it is literally Sisyphean. Modern physicists misunderstand reality, they make the mistake of thinking that because the field of force reality has no quantity that it isn't real. What they call reality is made from what the field is made of and not the reverse. These physicists would do well to go back into their history and study Plato who disproved Greek Atomism more than two thousand years ago. Before Socrates and Plato were the Milesian Philosophers who were also very much so empirical scientists and master mathematicians. Famous among the pre-Socratics was the philosopher Democritus and Leucippus. All this seemingly boring and academic history has a point to it. Mankind has travelled down this blind alley of atomistic, quanta, before but people do not know their history and therefore we are relieving errors over and over! Modern physics is in error and is not inventive, productive, and creative and has no achievements to point to that are of any use to the world.

Modern physicists argue and refute and mock the foundational understanding of a field with various pressures and entanglements. Modern physicists argue and mock Tesla, Charles Prodeus Steinmetz, Neil's Boor, Michael Faraday, Galvani, Benjamin Franklin, and Oliver Heaviside. These physicists contradict the real experimenters concerning the Ether Field as "disproven". So who should you listen to and side with on this issue? Do something! Think, reason, and realize, the men and their working understanding, rooted in natural philosophy of the past, gave you EVERYTHING which you have, use and depend on daily as the instruments of the modern world! The modern mockers have produced nothing real and useful by comparison. Who is closer to the essence of reality? Think!

The Ether is the platform out of which all that we call matter and the laws of physics come from! The Ether equals bands, lines of force in an endless, field of NO FINITE QUANTITY. At some measureable, quantifiable "time", traceable backward by extrapolation and rate of calculation, the lines of force acted. The lines of force, arguably demonstrated intelligent and conscious planning and purpose when they contracted themselves, from a complete state of rest. For no other discernable reason, the field warped itself or contracted itself thus becoming something more than pure inertial potential only. The field constricted inward, collapsed in a spherical configuration. This is the ancient Greek understanding of "Sacred Geometry" and Plato's understanding of the Platonic solids is built upon this understanding which all modern scientists agree these configurations are nature's actual manifestations. Human science is still very primitive; we could easily assume and insert the notion of God at this point asserting that God constricted the lines of force which make up the field. It is a human tendency to stop enquiry at a satisfaction point of knowledge and then to grow lazy and just simply say at this point we say God exists and makes. This is not an honest methodology, humans are afraid to look behind the curtain of mystery at different fixed points and we are all guilty of it. Why? Simple, we are afraid of what we will find out! We fear infinity!

Instead of stopping logical thinking and reasoning at an arbitrary point we need to go down the rabbit hole, which is terrifying. Electrical and Energy combustion research has given us some strange things to ponder. When voltage reaches extreme speed and voltage and force it simply disappears at some point in the cycle or circuit of power. This happens with combustion in car engines as well, the energy simply stops existing, it goes away, it seems to temporarily vanish, or does it? This cycle is known as the Hysteresis cycle. In the face of discussing the behavior of the field when it contracted, some may ask why after eternity did the field act suddenly and contract spherically down to a point? Based on high voltage electrical behavior we may be asking the wrong question. We are assuming the field as we nearly perceive and understand it is all there is. This may be a very big mistake. We may be dealing with higher and lower strings and dimensions within the Ether which has massive implications. The Ether has been illustrated as strings, a magnetic line and a dielectric line running in opposition to the other at a similar thickness or dimension with other similar lines woven 90 degrees juxtaposed to them. Consider that those lines of force may be just what we can perceive and utilize in this dimension but when the energy output exceeds a certain threshold we experience "black holes" and hysteresis! The nature of nature, the universe may possibly be analogous to the infinite strings of a guitar. Take an ordinary guitar, look at its width and vibration range, that string we call the "E". Descending down from the E we have A, G, C, D and then we have another E. The Ether may be just like this, when we send extremely high voltage down a circuit and vanishes it may be moving out of linear, x, y axis time and may move and shift into the dimension represented by the z line representing a higher faster energy dimension or string/line of force. When the energy is stored there it is temporary as it loses energy and then snaps back to it's default source and dimension. Light teaches us that higher energy light exists than what we can perceive and that it is smaller in wavelength, so it may be with the lines of force in general. Odd as it may seem, the analogy of a guitar may very accurately mirror our reality with the our known states of matter known as Solid, Liquid, Gas, Plasma, and Ether may very well correspond to the E Major, A, G, C, D with the E Minor representing the higher frequency line and crossover to a higher frequency dimension. For all we know the dimensions may be come in pairs and move in response to one another in an imbalanced ratio of phi opposed to 1.

So think of this, what does this analysis of higher frequency dimensions have to do with God and the contracting of the lines of force? We have to ask a new question and explore its implications. In other words we have been saying that perhaps "God" constricted the lines of force inward spherically so they would snap back in contraction to their rest thus tangling and forcing out in a hollow sphere configuration (The planets and stars reflect this formation pattern of the universe as they too are hollow). The snapping back to the rest state incidentally caused line entanglement in different configurations giving us the products for the states of matter i.e. protons, neutrons, and possibly what we call electrons and protons as well. To be sure those "Particles" and their designation represent a great lack of information and proper understanding of what they are and what they are composed of. We have assumed one dimension and therefore a God, or consciousness which caused motion/time. When we understand there are other dimensional configurations at higher and lower frequencies in the Ether we may be forced to ask a bigger question... "Did consciousness cause motion or did motion cause consciousness?" What does this question mean? It means if the resting, static field of magnetic and dielectric lines moved suddenly from rest the prime mover may have been a wave function come down activated by eternal and endless dimensional reaction, in turn, as the endless holographic field moves in on dimensions until it unbalances the descending dimension due to a phi to one ratio imbalance falling on descending dimensions!

The conclusions just drawn may be very reasonable when we consider what studies in light holography have unintentionally demonstrated to us. Dimensional descent may be very, very real as demonstrated when one gets a glass holographic portrait with a 3D image laser etched in it and then snaps off a small corner of the glass plate. When a laser is shined through the broken fragment of the hologram, the whole image is seen full and in tack, scaled down from the original. This weird property of the hologram continues recursively at infinitum, when a further piece is broken off the first fractured piece, it too contains the full image in a scaled down version from the first fractal. This is amazing, and possibly very educational and telling. Nature is teaching us but we are not listening. The rabbit hole of reality's nature may go way deeper than our fearful little minds can comprehend or *accept*. Maybe reality needs to be accepted as it vs. being "understood" as that represents fixed boundaries and there may literally be none!

The nature of holograms reveals a recursive and Sisyphean field of lines of force which hides the levels of the field and of all reality.

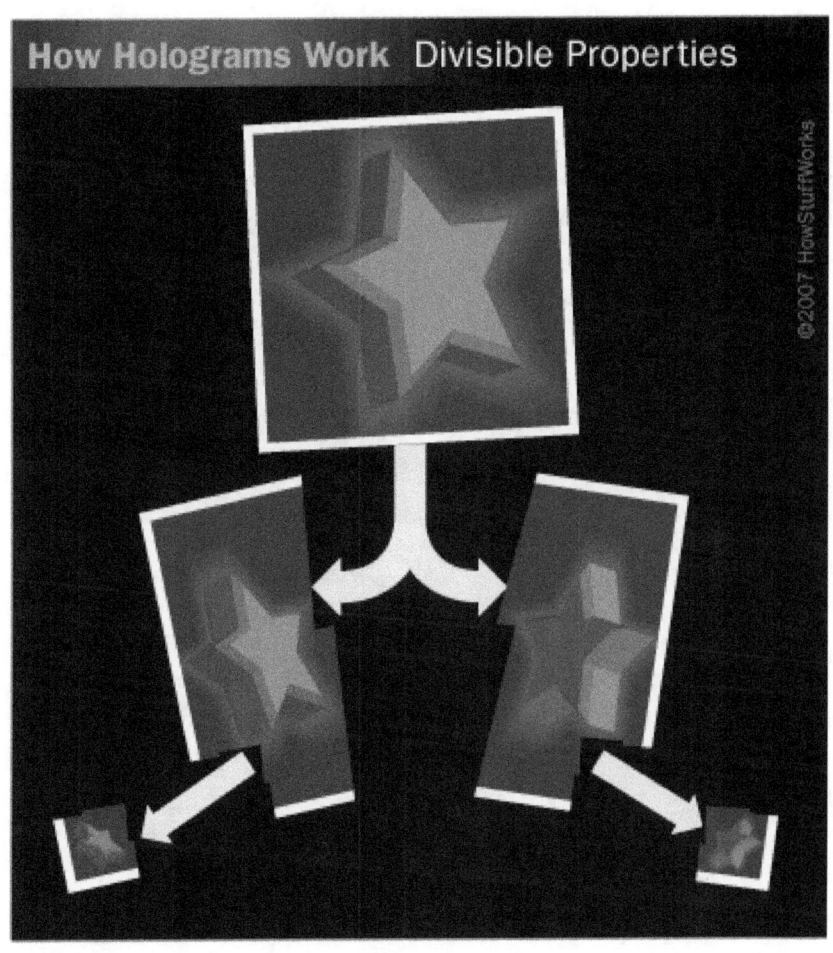

Example of a how a hologram is literally unlimited in how many times you can divide it down and still preserve the whole scaled down image, this is how reality scales down in an infinite recursive manner! Image found at http://how-does-things-work.blogspot.com/2010/02/working-of-hologram.html

Two More Faces and Names You Should Know Regarding the Recovery of True Science

 &

Doctor Eric P. Dollard   Ken Wheeler (Natural Philosopher

Eric Dollard clarified and demonstrated that electricity is NOT a product of electrons flowing through wires but along with Ken Wheeler demonstrates that "electricity" is a dielectric and magnetic phenomenon guided down wires, down "guide wires" to transfer pressure transferences through the Aether or Field.

CHAPTER 2

Eternal Lines of Force, the Field Which Contracts and Bangs in a Big Way

The observation and proofs, of the eternal dielectric and magnetic lines of force, are by far the greatest contribution to the electrical sciences made by James Clerk Maxwell, the father of modern day Electrical Engineering! These lines of force are what the whole body of content and calculations of the Maxwell Field Equations are even dealing with! Some of these equations are otherwise "adopted" by Albert Einstein and utilized for other purposes in his *general* and *special* theories of relativity, such as your famous E= MC squared equation. This equation is from Maxwell and is employed in measuring, tracking and calculating electrical behaviors and ratios, behaviors of the field. The field is much like Empedocles opined it to be as a center of evenly proportioned opposed forces, moved by tension, to act in conjugate to one another generating an incredible action. The two opposing forces are literally everywhere in the universe. The two forces he identified as "Love" and "Strife". Empedocles identified the two forces which we later came to call "Magnetic" (Electrical symbol *phi* Φ) and "Dielectric" symbol *psi* ψ) in the Steinmetz' and Dollard's Electrification Equations!

The Eternal lines of force are undisputed by all physicists. The majority of these same scientists, in overwhelming numbers, openly mock any mention of the Aether, never understanding that the lines of the field, and the Aether, are one and the same thing! The same scientists, beginning with Albert Einstein in his general and special Theory of Relativity, interpret the Michaelson and Morley experiment and others as proof that the "Aether" doesn't existed. What Einstein and his sycophants failed to understand, or at least failed to communicate to a gullible public, was that defining what the Aether is, was a dishonest matter of picking one of the many understandings and explanations of the Aether, and then knocking that down. The version which scientist Michaelson and Morley was not at all the Aether which Tesla, Thompson, Steinmetz were experimenting in. We need to step back and clarify reality's definition of the Aether. The Aether and the Field of never ending paired lines of force are one and the same! The field's magnetic lines of force are readily seen when iron filings are sprinkled on a piece of paper which is set over a bar magnet underneath. When two magnets are placed beneath a paper and are oriented in a north to north-pole facing position, the effect is that the lines of force are visibly seen pulling inward. The lines respond like elastic bands or springs, crushing in/crowding in together to push out and away from each other with force. The reader is strongly encouraged to perform this very same experiment on his or her own. This is 100% repeatable science experimentation!

The lines of force were first really noticed and experimented with heavily by James Clerk Maxwell of electrical engineering legendary fame. This is the same Maxwell who formulated the famous Maxwell equations of electricity which are heavily relied upon by all electrical engineers who make our electrical power distribution grid real and useful!

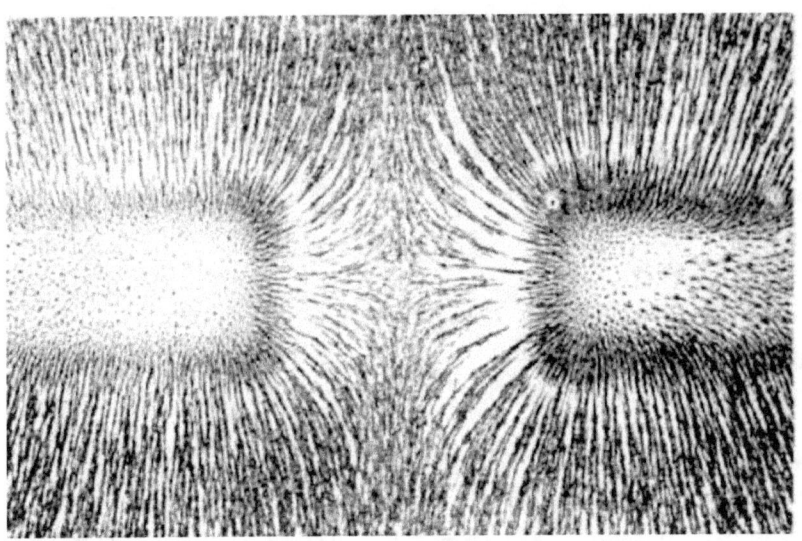

Chapter 2, figure 1. Carefully notice the lines of force are all around the magnets as revealed by iron filings. Notice the size variations. Lines of force come in an infinite size range, descending upward and downward with no beginning and no end point, just like fractions and fractals are asymptotic, recursive and Sisyphean.

In the image that follows the lines of force can be seen voiding the area between two magnets which are in a north to south-pole orientation. The lines of force are amazing and clearly real, at work, elastic in nature and omnipresent. You are looking at the substratum of reality! You are looking at the equivalent of a photographic negative of what "solid" atoms and atomic particles are made out of. Think of all the terms associated with atoms like electrons, protons, neutrons, neutrinos, gluons, muons, tack ions, quanta, even all of string theory, light, photons, electro-magnetic radiation. All of these are manifestations of different interactions of different tangles and configuration ratios of these unbounded lines of force!

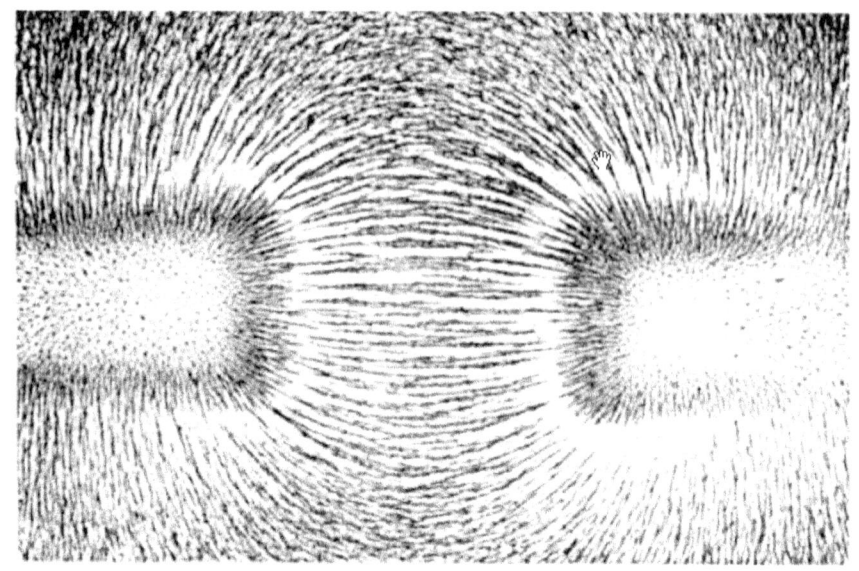

Chapter 2, figure 2.

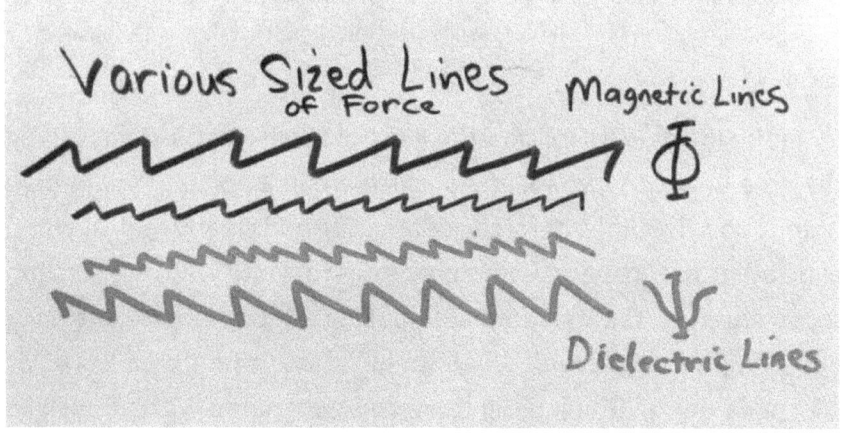

Various Sized Lines of Force   Magnetic Lines

Dielectric Lines

Chapter 2, figure 3, a closer view of the physical look of the lines of force.

The Field, what is it? What does it "look" like? What is its shape and configuration? First off the field is almost completely "invisible" in that most of it is unperceivable to the eye. Radio waves, microwaves, infrared, all of these forms of light are not visible to us but are completely real and present in our environment. These forms of light are just a portion of the "light" which is invisible but are manifestations of a disturbance in the endless field of lines of force. We call the different size vibrations different types of Electro-magnetic radiation, found along an endless spectrum which scales both up and down infinitely. To visualize the field think of grid paper. Grid paper is two dimensional but our grid is truly three dimensional (an x, y, z axis). This is exactly why math so precisely predicts physics and is the language of science. Think of a fabric weave of over and under warp and woof configuration as seen on a tennis racket. Now think of the same weave but also in three dimensions, like a Rubik's cube. This is a beginning point of understanding the configuration of the Field made up of paired lines of force.

Visualize the detail, now that you have a visual of the Rubik's cube, picture lines inside the individual box lines, proportionately smaller to the box lines. The lines inside the box lines are smaller to the lines of the bigger box by a descending ratio of phi to 1. Phi = a ratio like pi. Phi = 1.618034. Phi is the size of the lines of the individual boxes formed within our Rubik's cube. Inside of them are lines forming boxes which are size one (1) compared to the containing box in which they are found. Now the internal box within, the box has smaller potential light lines which are a one (1) compared to their larger containing box again. This pattern keeps repeating downward infinitely! This reality is demonstrated in the incommensurability of laser generated holograms etched in glass. When a light is shined on a laser hologram picture at the correct angle a three dimensional image appears. When ¾ of the glass is blacked out and a light is shone through the uncovered portion of the hologram glass image the whole larger image appears again, only in a scaled down version. All the information of the image on the larger plate is retained and permanent in self replication scaled down in size. The splitting of the image again, the separation of the ¼ of the image can also have ¼ taken off of that piece and it too will have all the information in miniature at infinitum of the whole hologram again. Light is telling us about the field.

The field is infinitely large and infinitely small and what was enacted in one dimension of scale was enacted and etched on all dimensions of scale. As light is imposed upon in film it also imposes on film and other "matter" its compositions and patterns. All matter bending light is itself bent light in a coagulated form and is conformed at the molecular level around light's matrices! In other words, the shape of the DNA strand and all that it can manifest through evolution is already patterned in the DNA as it is a reflection of all the dimensional potentials of the lines of forces when they are activated in the field to act as "light" (a.k.a. E.M. radiation). To spell it out consciousness and intelligence are imbedded in the light, or E.M. radiation, and manifest as driven to, or directed to, a destination by the environments in which living things evolve to better survive and gather food for species perpetuation. "Light", or E.M. radiation, exists in the larger field as a minority. Light, and all E.M. radiation forms, radio waves, x-rays etc. did not exist in our scaled dimensions, and neither did "matter" exist before the big bang. It is the "big bang" that entangled the lines of force to varying degrees to which gave us "matter" formed out of the entangled lines and harmonious paired vibrations of the lines which we consider E.M. radiation. These were the products of the disturbed field. The Big Bang marked the beginning of finite, quantifiable universe with all its finite parts and Entropy with ultimate heat death. Our field, as visually described, sat undisturbed and at rest for a beginning-less eternity as different sized paired lines of magnetic lines of force and their mirror image static (dielectric) lines of force, were set in motion from imbalance. AT REST, these lines were at low resting tension. Michael Faraday couldn't see it, he predicted the Aether as being configured as a cellular structure. This is an easy mistake to make as the field bends around the Farris metal to distort it's at rest state.

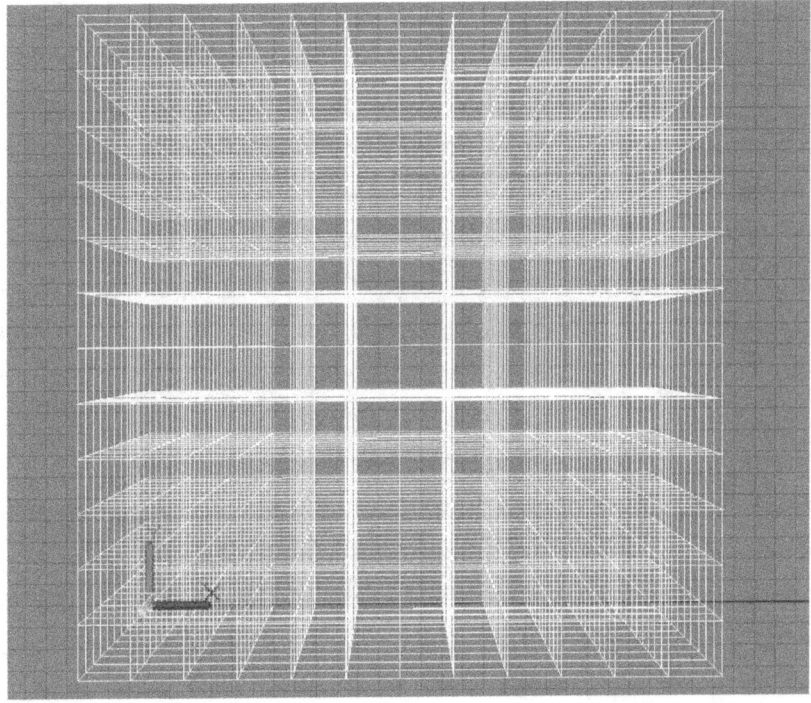

Chapter 2, fig. 4, the lines of force in the field of infinity. This model opposes the one commonly espoused by the poorly observed version of the field as set forth by the modern physicists. The Field is an x, y, z coordinate plane and the big bang expansion is a unit circle, or unit sphere, to express this all mathematically.

Thanks to Rene Descartes who gave us the coordinate plane which is also recursive downward, upward and is Sisyphean and asymptotic.

The graphic of the field can be thought of a graphic which roughly outlines the idea of the field but in truth the study and understanding of the field is still only in its infancy. The study of the field was interrupted and has not been fully rejoined. Think of the graphic of the field in fig. 2 but realize that within the boxes are down sized smaller versions of the lines that descend in pairs. The field is a field of fields, fields within fields within descending fields! These fields are whole chunks of lines of force which are separated from one another, ever descending smaller and smaller sets of lines of force. The smaller lines of force differ from their larger counterpart set by endless ratio descendants of *phi* to 1. This ratio difference is the cause of imbalance in the field. Imbalance in the field is what causes motion, movement. This leads us to just how a big bang was set in motion. Imbalance in the field could cause tumbling motion throughout the field but how could such motion start in a static at rest field? This is a key question. It can't! Yet there is motion due to imbalance. This suggests to my sense of logic is that the field, with its individual lines of force have to be composed of still smaller lines of force in set chunks. The lines of force in radio waves all the way down to sub-atomic sized lines of force are a separated, relatively isolated set in the field. Only when contacted by much higher energy from the upward tumbling smaller lines of force was the gap jumped which tumbled into our discernable set of lines of force. This leads us to how a "big bang" contraction and then expansion could occur in a static field. The imbalance of ratio is the discovered "Sacred Geometry"/Cause.

UNDERSTANDING THE BIG BANG. ANSWERRING, WHAT BANGED?

Light and matter are not the same thing but they are made out of the same stuff. Both light and "matter", atomic matter, are made from the interaction of paired lines of force (magnetic and dielectric) in various interplays and speeds. The lines of force, as they existed, pre-big bang, were acted upon by something Nicola Tesla discovered, THE ROTATING MAGNETIC FIELD. Tesla did not invent the rotating magnetic field, he discovered it in nature. What Tesla did was to mimic and capture and contain in a mechanical device, just what nature does when it creates lightning and electrical forces and discharges them. Benjamin Franklin did so similarly. When you see a strong magnet pull together to another very strong magnet you are witnessing the big bang on a small scale. The resting lines of force (magnetic and dielectric) in their lattice form were sucked into a *relative* center point quickly, like a giant magnet. When the field turned in resistance, it then behaved as electricity or magnetism always does, it repelled out just as two magnets when spun to face one north-pole in alignment with another north pole. In other words the field sucked in like a sink hole and then repelled super rapidly, (a.k.a. banged out). Unlike the big bang proposed by particle physicists of today, we understand by this analysis that the big bang was not just an explosion. The big bang did not just force out from a central compressed point, as what was compressed was spring like, or rubber band like lines of low tension. This means the bang both pushed and pulled the lines of force back to rest.

In every way this understanding and explanation of what banged is superior to those offered in modern physics. Modern physics tells us about a big bang which banged into existence all time matter and energy. Modern physics tell us about our entire universe banging into existence out of literally nothing. The man who gave us the big bang theory was content to leave us with this riddle unsolved as he himself was content as a Christian scientist. The big bang is one of many incomplete explanations of modern physicists who leave room for guessing, beliefs, superstitions and speculation. This is the point where physicists and religious people alike leave the understanding of what was before the big bang to the speculations that gods or a god created the universe, "the heavens and the earth" out of nothing. So the reluctant scientist as well as the religious default to a spirit creature of endless intellect and power a responsible for creating all. The scientist at times is more reluctant but ends up saying almost the same thing that matter just banged into existence from absolutely nothing at all. Something comes from nothing. They don't explain why this doesn't just happen again every ten minutes. This is not science! Even Aristotle, the earliest and greatest of our scientists, fell into the trap of laziness of intellect which defaults to a god when we get to the end of our observations. Plato also fell into this trap.

No physicists as our premiere scientists have failed us. They offer us no scientific explanations of anything before the big bang. Their model falls short to do this, their model is inadequate and cannot be compared to this understanding of the nature of light, the field, the lines of force etc. On the other hand, the field and lines of force science perfectly explains why the big bang happened, what matter, light and time are made from and why the bang happened at the time it did. The big bang happed in the field, the static resting field of magnetic and dielectric lines right on cue, in its turn. What does this mean? The resting field which we recognize as our own universe is the portion of the E.M. field which is in a ratio of 1.618034 compared to a paired set of lines which are in a ratio of 1 juxtaposed to our 1.618034 set of recognized lines. That set of lines was banged into by a smaller faster set of paired lines of higher energy acting as a magnet of spinning poles. This in turn was affected by a smaller faster set of lines which tumbled into them on and on as they descend into each other like dominos in turn with no start point and no end. Where did it start, it never started, it never ends, the domino effect is incommensurate! The motion is a product of imbalance just as if you had a level teeter totter which you put apples on. On both sides you put one apple of equal weight, they remain balanced. Put an additional .618034 part of an apple on one side only and the teeter totter goes into motion, so it is with the lines of force tumbling and rotating. The action is very much the same as a north side magnet being put near to the south side of another magnet, the sucking together happens, the collision tumbles the action of the higher energy set of lines to rotate to orient as north magnet meets another north magnet (our set of lines) causing a massive big entangling bang.

Chapter 2, fig. 5

Chapter 2, fig. 6

Chapter 2, fig. 7

We have answered our question of what the Field (The E.M. Field) is by composition and configuration and this leads us up to how matter and light and other E.M. spectrum realities came to have the forms we now have, know and use. The Bang manifested matter by creating what we perceive as "particles" in atoms. A second manifestation from the Bang was the, quite separate, E.M. Spectrum. The interaction and collision of the two manifestations of the bang manifested the living world with the environments which accommodated them! Because the wave that reach our discernable portion of the E.M. Field was coming from one direction means that "time" came into existence traveling in one direction. So we must explore the possibilities of time travel and how this relates to space travel.

"Time" does not exist! As Aristotle clarified, time does not exist in the sense of the past, present and future is all here together, they are not and this is absurd. Time does exist only when we understand that time is motion! Motion relative to other motion can be compared and measured with numeration next to one another; this is the real version of time. So no, it is not possible to go backward in time as the traveling wave through the E.M. Field is one directional, upending matter as it goes; manifesting life codes in short life spans. Time travel is not possible for the individual cell trapped bits of consciousness which we are in the material or matter manifest (atomic) universe. When the consciousness, found in the individuals, returns to the whole of consciousness this is impressed onto the all events patterned and written onto the E.M. Spectrum, the big DVD. In the E.M. Spectrum material movement and time is not the same limitation it is in the atomic world. There, time is not the same, there past, present are all there

95

simultaneously, the future is still there too, patterned and is therefore not there.

The traveling wave of metaphoric dominos (ascending lines of force) was and is set in motion because of a ratio which has left its impression and code on our creation. Both modern and ancient scientist and mathematicians have noticed that the universe, galaxies, stars and living things on earth all manifest on a reoccurring pattern. The pattern is called many things, the golden ratio, phi, the divine proportion, the Fibonacci sequence, a relation in nature close to what fractal geometry reveals. The golden ratio is 1.168034 but that is just a number until you compare it to any other arbitrary measurement of "one". Nature keeps repeating that motion, growth and activity keeps manifesting by juxtaposing the imbalance of the ratio of 1.618034 to 1. The interplay of this imbalance is the source of all growth and motion and manifestation over and over in nature. This is odd but real. It is by this fact that we know the lines of force descend infinitely in this same ratio scheme and in fact the reality which we witness is formed by their interplay.

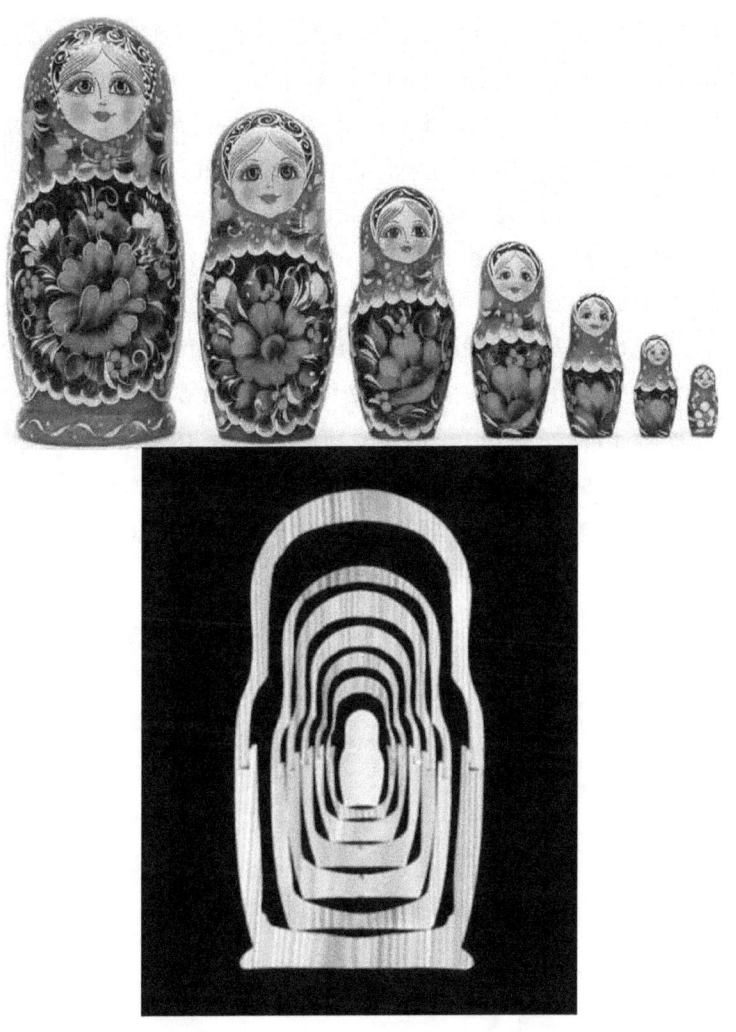

Chapter 2, fig. 8

Look at these Russian nesting dolls. For those unfamiliar with these dolls, realize that these dolls open up in the middle to fit next smaller doll into the opening over and over. The Field consisting of lines of force, is just like these dolls, directly within or descending within. The field consists of magnetic and dielectric paired lines descend down recursively in a ratio of phi to one at infinitum! The imbalance in size is triggered into motion in a tumbling descent and has always been tumbling upward into our detectable dimensions. Such eternal, boundless descending inward and downward is a mathematical reality as is demonstrated on computer models based on the equations of Begonia Mandelbrot in his discovery of FRACTAL GEOMETRY. Entertain yourself by going on YouTube and watch videos of Fractals endlessly focusing in on one area only to have focus in again and again and again never ending ratio discordancy based on the "Mandelbrot Set". Image of Mandelbrot and of Computer generated fractal geometric images to follow …

Benoi Mandelbrot Fig.9,

Figure 10.

A view of what the contraction of the lines is also a view of the expansion or explosion of the "big bang" in how it manifested. The

99

great contraction of the lines of force was downward and spiraling just like a drain of water spirals inward as a whirlpool. The expansion was the perfect reverse of the spiraling whirlpool, ultimately manifesting in a spherical form just as Aristotle predicted. The lines of force are of two mirror types just as Empedocles figured out over two thousand years ago. Empedocles divided the primal force into two opposing actions, which he called "love" and "strife". Empedocles explained that love pulled all things together but that strife pushed all things apart with explosive force! The nomenclature of Empedocles is poetic and therefore mocked and dismissed by later "scientists". They should have listened! The lines are of force are real, magnetic and dielectric and identified by the ancients and further Empedocles conceived of the big bang, two thousand years before Steven Hawking was born, as an action of contraction and expansion caused by the lines of force! Chew on that and realize the foolishness of modern physicists.

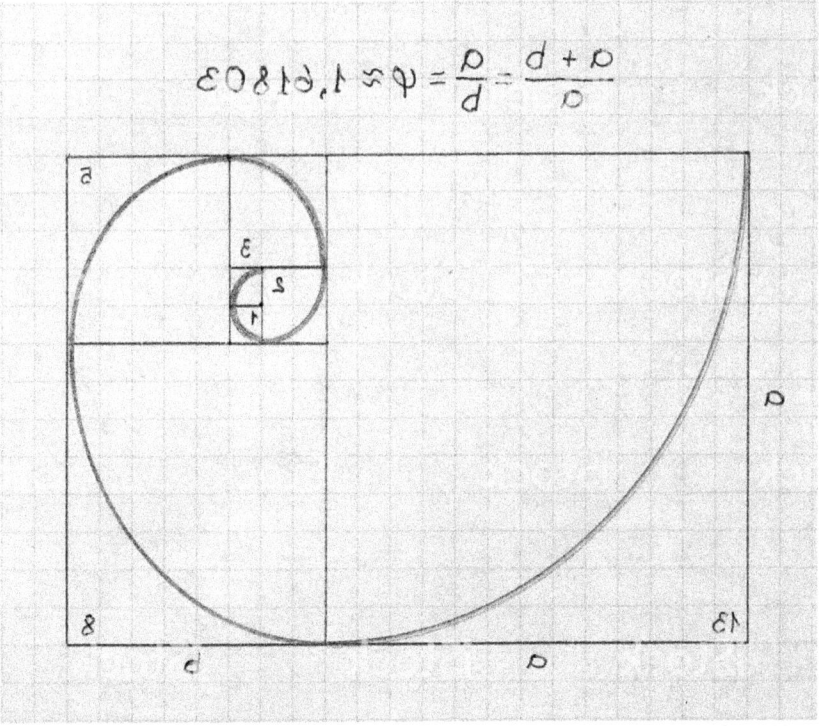

First the collapse in...

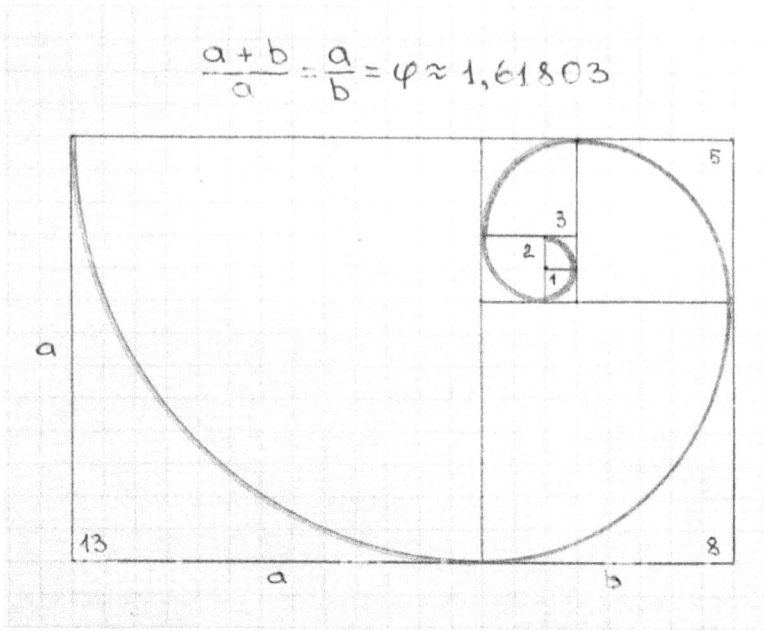

$$\frac{a+b}{a} = \frac{a}{b} = \varphi \approx 1{,}61803$$

Next the expansion out...

Fig. 11. This and the following graphic show just the pattern of expansion which was generated out of the lines of force first contracting according to the same pattern. As both Anaximander Empedocles conceived the primal substratum field existed and caused the known states of material reality which divided into the know states of matter of solid liquid and gas! Also their was know some concept of the material state known as plasma as fire too was considered another manifestation of the primal forces produced in the field. These proofs were derived through largely by logic and the math brought to us and Plato from Pythagoras. Their age of scientific discovery has not been properly respected by the modern physicist and this is their downfall. This is not the case for Steinmetz and Tesla, these intellectual geniuses are largely the cause of these discoveries as they headed the natural philosophy of the ancients!

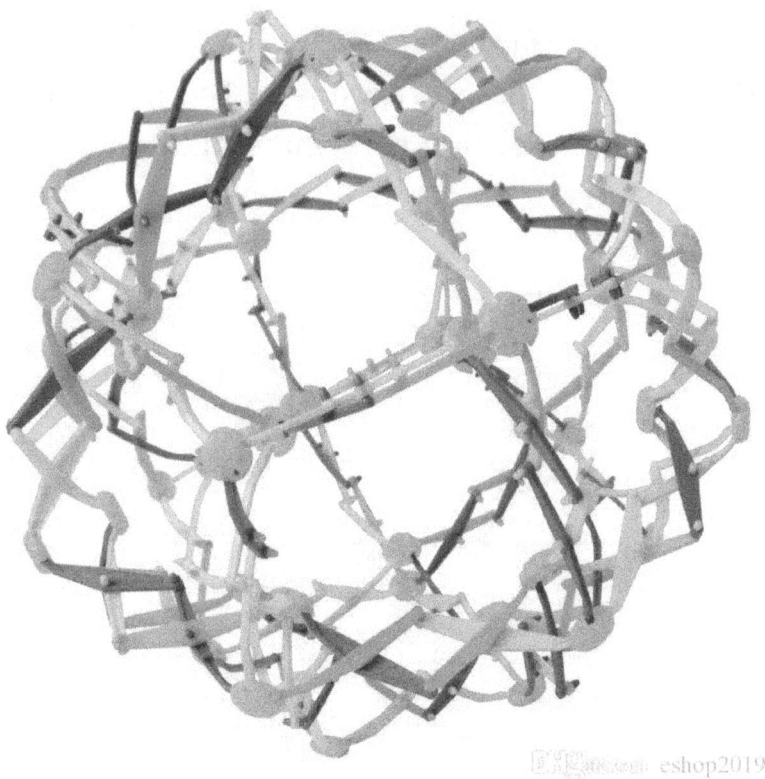

Figure 12. Lines of force, squeezed in a contraction. According to Plato and sacred geometry the first shape of material creation was a sphere. This was the Big Bang's expansion pattern by logical deduction. This is what Aristotle observed, that the universe is not infinite but a sphere inside of the substratum.

CHAPTER 3

Products of the Big Bang, Tangles Lines of Force. What Little Atomic Particles Are Made of, This is the Birth of What We Misunderstand as "Matter"

Chapter 3, image 1. J.J. Thompson

Atoms, what are they, they are what they have to be, let's approach this with observation plus logic. Above is the image of Sir Joseph John Thompson the discoverer of the electron by means of observations in Cathode Ray Tubes. What Thompson observed was seemingly the separation of parts of atoms. Before the time of Thompson there was no clear understanding of the components of Atoms, Atoms were thought of kind of like the ancient Greeks, Democritus and Leucippus conceived of them as solid with no void in them, completely indivisible. Along comes J.J. Thompson and demonstrates with luminous gasses in a tube that atoms are being separated by electrical charge, manifesting their polarizing opposite components, specifically the electrons. Instantly those of the physics community named the electron a "particle" and that is fine in that an electron, whatever its true form was distinguishable as a unit, a quantity of various charge sizes. This is useful and therefore a good development. There is a problem however that arises when we try to make everything in the universe conform to the atomist's model of reality or atoms moving in an empty void.

The problem of latching on to the idea of the ancient atomists is, doing so fails as a literal model to account for electrical phenomenon and electrical behavior through void space free from electrons and fails to explain magnetic action which involves no electron movement whatsoever plus it fails to account for blocked actions. In other words particle insistence fails to be able to explain how a magnet one foot away is moving a second cubical magnet when turned which is situated on the others side of a one inch thick plate of copper. Absolutely no linear electron flow is possible, the actions is occurring strictly because as Galvani observed there is an atmosphere (Field bending) around certain metals! No "particles" are involved in this action and no particles could be. Wireless transmission even occurs through certain densities of lead which is what is used to shield against high energy X-rays when people are being examined at the dentist for example. Why lead? Lead is "particle" dense, a heavy metal, the transmission of free electrons through it is almost completely retarded to a null point. Even with these facts set and indisputable in the real world you have "physicists" who stubbornly insist on the ancient Greek model of Atomism. Atomism on this model would make wireless transmission through led and through the vacuum of space quite impossible yet we do both every single day!

Now with this being divulged let it be stated again, quantification for the purpose of calculation is a good thing, a valuable thing, a useful thing, without it electrical and chemical engineering would be nearly impossible to do. It's okay to cut lines of force up into quantifiable units for calculations as long as we do not lose touch with reality, as long as we do not forget what we quantify as a particle or a wave is a point of compression or rarefaction on a real tangible dielectric sea in motion! Scientists like Albert Einstein and Steven Hawking are just two such scientists, physicists who next to nothing of electrical engineering nor did they use electrical systems behavior as their jumping off point before developing their quantum particle based theories of curved "space-time" and the "speed of light" and other such detached from reality based ideas. In doing this they have mislead millions of less informed minds, and disconnected them from the substratum of true science which unifies Electro-Magnetic Radiation Field behavior studies with the formation of DNA and life itself as it manifests on an consciously intelligent grid. By doing this Einstein and Hawking undermined their own goals of unifying all sciences as they cut and partition things that are all connected by lines of force. Think of just how self-contradicting quantum physicists are. How can you have a wave-particle duality? That is the same thing as saying you live at the north-pole and the south-pole... at the same time! These people play with Maxwell's equations and try to make them for use on their models and the <u>deeply</u> <u>ignorant</u> population calls the "brilliant geniuses". If some drunk walked up to you and said that he lived at both poles, both north and south, at the same time you would laugh at him and call him "stupid" but when it comes to these physicists who tell you they are brilliant because they cover their non-sense in other people's equation which were never intended as proofs of their non-sense, you call them "brilliant". This is commentary on your own intellect if you play along with, and celebrate non-sense.

Like Socrates, let's apply a Socratic break down of someone's claims. Let's play along, we are told that an electron, for example, is both a wave and a particle isolated in a vacuum at the same time; let's take that at face value to begin with. What is a wave? What do we ever see wave in nature? When we look at the side of a mountain on a windy day we see waves in tree tops, just like other waves. In a still pond we see no waves, until, we throw in a rock into the still pond, and then we see a suctioning down of the rock which generates a voidance in the serine medium of the water. Energy was input into the system and the system reacted to it and was set in rippling motion. The medium which was still is now waving. Waves are evidence of a medium waving! You don't see wind waves on the side of the mountain, you see trees waving, and grass waving and water waving. Okay this is logic breakdown part A. Now does any non-fluid or particulate ever wave in nature? Does a bowling ball wave in the wind? Does a rock wave in the wind? They do not.

How do we know something we can't see is a particle? We know by it making contact with break lines in between. We detect, as J.J. Thompson did, pings or intermittence contacts or registrations. Is a solid particle the only thing capable of registering pings or contact hits in all of nature? Can nothing else do that same exact action? Can a wave register pings? Do waves have peaks and valleys? Yes they do, in fact that is what defines a wave. Are not peaks and valleys in a fluid system's points of compression and rarefaction to the point of the compressions being a solid in contrast the rarefactions carried out in a liquid medium? Why yes they are! Hmmm so our proof that a wave particle duality is both a wave and a solid in a void seems to model a phenomenon never seen in nature at all. A wave in a tangible medium answers the reason of why do we sometimes perceive an electron or photon as both a wave and a particle. This is an easy experiment that a child could do to demonstrate how you can get solid readings of contact and waves from the same action. Fill a cake pan half way with water; put a mark with a permanent marker at the still water level. Next put a fan on the opposite end of the pan to blow on the water. When the fan begins to generate waves in the water mark a second line where the height of the waves is making contact with the side of the pan. Turn off the fan and note the difference of an inch or a half inch for example. Turn on the fan again and count every time a wave peak contacts the upper mark and you will have discovered the secret of the particle pings of J.J. Thompson and all the rest who found "Particle" proof by the same experiment only in a Cathode Ray tube and a medium of luminous gasses!

Good old Socrates you solved another mystery which will probably upset those who are married to the idea of particles in a void. Your logic helps us resolve the dilemma of the non-sense "wave particle duality" observed, only this explanation is inclusive of all phenomenon observed as to where sticking to a hard line of atomism fails to account for electrical phenomenon side by side with alleged wave particle duality. This is the same problem Einstein ran into in his special and general theory of relativity when he tried to explain, with particle physics, what he called "spooky action at a distance" which he observed. This is the subject of the next chapter.

The Error of Charged Particle Notation

Out of all this phantom chasing of "particles" came the notion of "charged" particles, particles which are "electrical", which have either a "positive" and "negative" charge allegedly based off the work of Benjamin Franklin. This is comical, Benjamin Franklin meant no such thing as charged particles when he made up the system of notation denoting positive and negative in regards to electrical forces.

The origins of your concepts of positive and negative in electricity are not what you have been taught in conventional physics, chemistry and science. The idea of charged particles is completely arbitrary and has nothing to do with "charged particles" as you may have been misinformed. You see, the origin of the idea of designating particles as "charged" is completely arbitrary and had nothing do with polar opposites as it has evolved into through misinformation. Benjamin Franklin is the person who invented and conceived of the concept of describing electrical behavior with positive and negative designations. Those who followed completely misunderstood and confused Franklin's simple meaning of using these designations. The ignorance of what Franklin meant by his use of positive and negative designations in electricity, added to people's ignorance of electricity and set back the research one hundred years!!! I cannot overstate this, please understand, electricity is far simpler than what people have made it out to be. Fools have taught other fools, that have taught vulnerable students, that electrons are "negatively" charged "particles" in opposition to positively charged particles. Franklin, who invented the idea and designation was an accountant and printing press man who was simply using positive and negative as a system of ACCOUNTING of quantities.

Was Franklin "Wrong" in his Designations of "Positive" and "Negative" in Electrify as Modern Arrogant Physicists Teach Us?

To elaborate and to clarify, based on Benjamin Franklin's own writings, franklin was simply using positive and negative notations to track the movement of electrical fire levels and deficits. Allow me to clarify, Benjamin Franklin, ACCURATELY observed, that electrical energy was not being created by friction but it was simply being moved from one, as in being removed from one area and that same quantity was being accumulated in an opposite location by contact. One location had more, too much electrical energy by taking the same exact proportion of it from the losing side and adding it to the gaining side. At rest to opposite pole locations are not charged DIFFERENTLY until by mechanical or chemical action (as in a battery) one pole has for example 5 electrical units stripped from it, that is the exact amount of electrical units which will be added to the other pole in the closed circuit (circle) of electrical energy. It is that simple! This is all that Franklin was communicating when he used "positive" and "negative" notation when describing electricity, nothing more and nothing less! The simple concepts of anode and cathode simply designate quantitative proportionate movement from one electrode to another electrode.

With the rise of the physicists came the rise of all kinds of disinformation, misinformation and political bias. Physicists loved positive and negative designation for their allegedly polarized oppositely charged atomic particles which they imagined floating in a nothing of a void. None of that is true nor accurate nor what the original inventors of the designations and particles ever believed nor intended. Their discoveries and breakthroughs in understanding were appropriated by those of the physics discipline to hijack and lock up in mysticism, confusion and layers of contradictory terminology to HIDE their meanings. Why would anyone do that? I won't speak to their motives, you just need to know that this happened and all these facts are easily verified.

So back to the fuller topic of this chapter, we learned what atoms are NOT. We also learned that what is now considered the meaning of charge is NOT what was being noted by Benjamin Franklin. Now it is time to follow through and figure out what atoms ARE. Atoms and particles of atoms are manifestations of, discernable in the medium, pings, contact points. The parts of the atoms are not positive and negative so much as they are opposite and conjugate, different in direction, pull, motion and goal. The lines were squeezed together about <u>13.8 billion</u> years ago to a stress point so intense that it blew out, exploded as well as snapped back out to their original resting points. This action was violent and energetic beyond our comprehension but the evidence is plain in the universe. Background radiation levels testify of it, the much higher blue shifting stars to red shifting star ratio from our vantage point testifies of a huge explosion trackable by calculation to our 13.8 billion year mark for the event. Now put the lines of force, the field back into position which the physicists removed from our model of reality and what do you get? Test it for yourself, do this small experiment, go to Walmart into the fabric department, buy several spools of white elastic band and black elastic bands. Also buy bands of different thickness and dimension. Tie the bands to a chair in alternating black and white series, next walk with the bands in your hand as far as resistance will allow and then release the bands to snap with force in the direction of the chair. Go and examine your results. Your results will be the same as seen in the big bang. You will discover white lines tangled with other white lines (Equals a negative particle). You will discover black lines tangled with other white lines (Equaling a positive "particle or relative solid point for example). You will find other combinations, neutral or neutron combinations, lines knotted on themselves. You will find combinations of black and white lines of the same size tied together (Nucleus' of atoms). You will find tangles dominated with more of one color or size of line (Representing your ions and isotopes). It is a very good exemplary model experiment of the big bang and the products produced by it in the lines. These examples of tangles represent what we consider solid and particulate matter but another different product comes out of the experiment.

Electro-Magnetic Radiation Both Large and Small

Initially when you pulled the dozens of lines to their maximum point you had them organized in alternating order and alternating sizes. When you snapped the lines some did not tangle at all and others only twisted together loosely or braided together. These manifestations represent something that is not solid particulate matter at all, this is "light" in all its form or more correctly, electro-magnetic radiation. E.M. Radiation comes in all different sizes from as large as a mountain to smaller than the nucleus of an atom but they all configure the same way. A conjugate, juxtaposed braid and dance which later when pushed through atoms gather the electrically formed nucleotides to graph onto its superstructure to later manifest the generic DNA of life which can manifest endless combinations of animals and plants as prepared to the environmental requirements. The environmental predispositions are in the structure of the Radiation and manifest in the material world in match.

This is an example of a simple particle or atom.

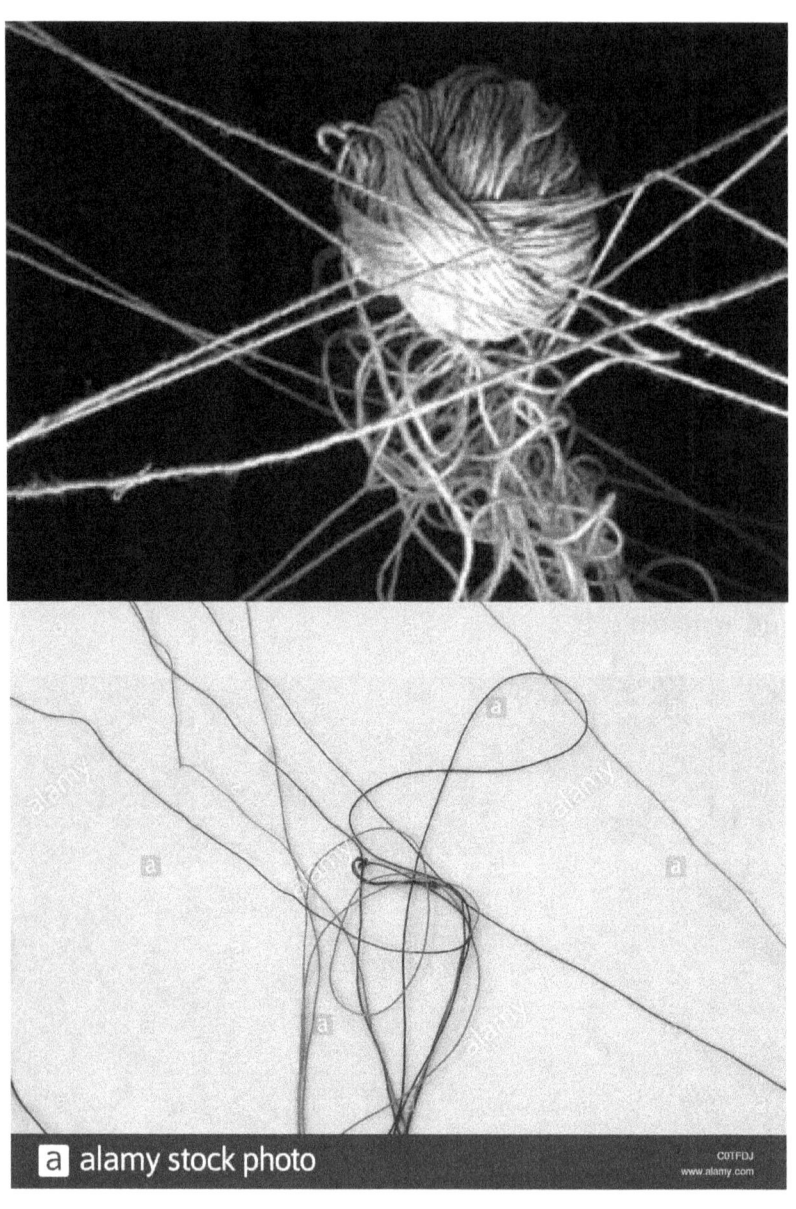

CHAPTER 4

Explaining Einstein's and Sir Isaac Newton's "Spooky", Instantaneous, Action at a Distance (IAAD) Which he Couldn't Explain With His Quantum Physics

Albert Einstein did not believe nor understand that the field which I have been describing so far. Einstein was the poster child of modern "physics" or quantum mechanics which he depended upon to understand, describe and predict the behavior of nature at its most fundamental level.

Trumping Einstein begins with us looking deep down into atomic structures. Physicists inform us that atoms are really, really tiny, and that the nucleus, composed of Protons and Neutrons are orbited by "electrons". We are further told that these atoms consist of mostly "empty space". Numbers like 99.999% of atoms are empty space. This of course is what they say to the public, when studying the subject up close in greater detail, the physicists will concede that their precious model of particles floating in an empty void vacuum are not technically accurate. Quantum physicists grudgingly admit that the so called empty atom is in fact full, loaded with tiny E.M. fields.

Quantum physicists, will recognize Einstein's quotes about a phenomenon which he could not explain with his quantum models of reality concerning "Spooky, action at a distance". Einstein observed, as did Newton that there are movements or actions of mater in one place which are clearly not acted upon any line of particle to particle interaction or contact. Particle, quantum physicists seem content to live with this alleged paradox/anomaly instead of going back to the drawing board and to do experimentation to establish the true nature of reality. This is what is different about the scientists prior to Einstein, the "Electrical Scientists". There are really no more electrical scientists and experimenters. The difference is that the electrical scientists were research and empirically scientifically based and they produced the real products of electricity which we use daily! The big difference between the modern physicists and the electrical scientists was their understanding of what they called the "ether". Physicists explained away the ether as to where the electrical scientists worked with it understanding that it was real, it was the field of electromagnetic forces and how they combine to form plasma and electricity.

WHAT IS TIME?

Time is nothing more than motion and motion has two ingredients, shape and imbalance. In the field there are parallel lines of directional bent, think of the images of lightening, on one side of the bold we have magnetic lines pointing up while on the opposing side we have static lines of force pointing down. These paired lines are motion and creative motion when tipped into imbalance. Observe the image below, it appears to be in motion, but actual, motion causes entangled lines which cause matter.

The Field is conjugate in configuration (If you could see it as a whole) like this picture above sits in potential and implies motion according to perception and directionality which is activated in turn by ratio differential which exists in the difference of line pairs compared to one another in a phi to 1 difference or ratio. Motion is shape and perception & perception is motion.

The lines of force are paired with equally sized lines such as the illustration above shows. Motion is implied by shape. For example the lowest horizontal row implies a motion to the right as to where the row horizontal line above that is implied to be in motion to the left. Stare at this picture it is what your eye and brain are signifying even though the image is actually not changing positions. In the E.M. Field the lines are similarly shaped and predetermined to imply movement juxtaposed to its fellow paired line of force. Unlike the picture above the line pairs are distinct and also are smaller to one another in pairs in a ratio of 1.618034 to 1 or phi to 1. The pairs, because they are uneven in size will be set in motion in vibration turn like the strings of a guitar which is strummed from the bottom string up to the top string. See the Graphic...

GUITAR STRINGS IMAGE

Entangled lines can have more than one knot in them, Einstein's I.A.A.D... or spooky action is no more difficult to explain that knots in a string. Tie a string to a door knob; put a visible knot in the string about two feet down the line. Tie another knot in the same string about four feet down the line. Tie the opposite end of the string to a chair with very little slack. Now start pulling the string toward you from the door knob side, put tension and pull on the string. This is what gravity does, this is what electricity does and so when you pull on the string, tense it, it pulls not just the knot closest to you in, it also pulls in and moves the knot four foot away from you! In some respects Einstein was not much of an "Einstein".

Albert Einstein and His Special and General Theory of Relativity under the Microscope

Albert Einstein did many things right, he made good observations and employed (Plugged in) trusted and true math equations to his theory, the Maxwell Field Equations and Euclidian Geometry as well as Pythagoras's Theorem. $E = MC$ squared is part of a larger equation derived from James Clerk Maxwell. The Problem with Einstein's work is that he set back empirical science and logic hundreds of years by redirecting the science and logic of brilliant men who came before him as I will demonstrate. Einstein's math is solid, but it is applied by the afore-mentioned great thinkers and scientists to further uncover the mysteries of what they ALL CALLED THE AETHER or the FIELD. Einstein did not like the science of the Aether or the Field and by mischaracterizing the Field found "Proof" to dismiss its very existence. Einstein misunderstood what the field is and does and what it is not and what it is not connected to. I do not know if his mischaracterization of the Field was deliberate or not, no one does. Einstein short circuited his own proper understanding of the field as his love of and marriage to particle physics did not allow the field to be the Aether. The same bent also ruined Einstein's ability to focus on the electro-magnetic field as the material substratum, the source of detectable matter.

This begins Einstein's massive failure which the failure sycophant science community never figured out. Failure to recognize the lines of force as an atomic substratum source is the genesis of need to generate a new particle based physics with many, many patches and contradictions. "Particles" are charged electrically, "positive and negative". Also in particle physics logic we also have atoms which are more positive than negative or more negative than positive, based on a ratio of negative to positive particles, these are what we call "ions". This variation is a ratio involving the particles which particle physicists call "protons" and "electrons in varying ratios. Anoter particle which they calculate in radioactive elements is called a "neutron". How many extra neutrons an atom has puts it in the category of an "isotope". This is a construct system, constructed in order to measure, calculate and manipulate for use of atomic energies by dividing them into useful quantities. With this stated, it does not mean that protons, electrons nor neutrons are actual "particles" floating in a void/space. It is this understanding where physics took a hard left turn which had to happen as particles in a void by definition cannot affect one another at a distance until and unless they collide into one another. This is where Einstein's own words testify against the value of his theory as he observed "spooky action at a distance" occurring with distant particles moving in concert when localized particles in the laboratory were effected as if paired, distance notwithstanding. Particles at great distances were moved when localized particles were moved in parallel opposite, mirrored directions. Particles in a vacuum, of course cannot move that fast, instantaneous regardless of distance, and this became a monkey wrench to the theory of relativity. Einstein had already declared that nothing moves faster than the so called "speed of light". Instead of admitting that the theory was overthrown by the failure to explain this phenomenon Einstein doubled down and created a patch for his sinking theory. Einstein brought back the Aether under a new name, gave it the properties it had as the Field to move and mirror paired particles by means of the fabric of a "space-time" fabric or continuum. Very clever! Space, which is not a thing, now had properties and curves which explain away gravity, magnetism and instant spooky actions at a distance by particles. With this Field which is not the old field, not the old Aether but does what it did, the maximum of a speed limit of light was navigated around. The need for colliding particles was also navigated around as well by this new fabric or medium... Hmmm, the new Aether is not Aether, the new field isn't a field but does exactly what the field does. Brilliant bait and switch, the art of a master but it is a counterfeit and therefore a pale imitation which does not account for many other factors without further patch explanations. By switching Aether for Non-Aether fabric Einstein displaced all the old masters of science and electricity by this action and set himself as the master of the new physics because he is the author of the new terms. Did you get it? Did you catch it? No delay in "time" equations can explain the instantaneous action at a distance by the action of colliding particles because I.A.A.D. is happening much faster than "photons" can cause a chain reaction of speed of light relations at a distance. Einstein could not explain this without a *new* Aether. In this Aether "spin" and "counter-spin" can affect distant particles having the same action and mirror action in the wake of this "quantum fluid" called "time-space".

So many contradictions keep poking holes in the sinking ship which is Einstein's Special & General Theory of Relativity. Einstein, instead of confessing that he was wrong about the field covered his tracks by making a patch. After dispatching the Aether Einstein found that he needed it back to explain I.A.A.D as only a field of connected elastic lines of force, conjugate and omnipresent could explain I.A.A.D. Einstein needed something just like the Field to move distantly paired particles instantaneously, faster than the speed of light so he reified the non-real things such as time and space to play the role of the conjugate fabric making such action possible. Never mind that particle physics completely contradicts these possibilities, time space fabric and particles in an empty void cannot be reconciled they are not mutually inclusive models of reality as by logic they are mutually exclusive. Modern particle Physicists based on Einstein try to explain I.A.A.D. with the concept of "spin" if one particle on this side of the universe spins right, the counterpart on the other side must spin left in mirror action. But they say this happens through a nothing, void of space, which they turn around and say is a fabric (like and Aether) without saying Aether, they call it "space-time" to replace the real conjugate field with an imposter cheap counterfeit. Sometimes because they need the medium they might call this property of "space-time" something like a quantum fluid action. Sounds exactly like Aether. Nice try.

Einstein should have realized that you can't simply replace or rename the Field/Aether the "time-space continuum" now because Maxwell and Hurtz proved, the field is composed of conjugate lines of force which are Magnetic and Electric, not time and space. Maxwell, Hurtz, Plank, Heaviside all proved that the field, Aether, with its magnetic and dielectric lines braided together produce electricity and I.A.A.D. by manipulation of these specific lines! Einstein's chain reaction of failures in logic and empirical observations found in his Special and General Theory of Relativity was all caused by the false premise that atoms are composed of floating, independent, particles. What are particles anyway and why do they seem to be "charged" with charges? Relativity has no answer for this. No answer to this is even possible with the Theory of Relativity, so stop clinging to it! Electrical theory and research was destroyed by the Theory of relativity as it had to declare incidents of magnetism and gravity were merely features of curved "space" and that the conjugate lines of force don't even exist. The theory reduced these lines of force to mere streams of electrons guided through wires. This is exactly why electrical advancement stopped in the 1920's. Einstein dispatched the truth of the field, the only explanation which perfectly answers and accounts for the existence and creation of matter – PRIOR TO THE BIG BANG happened. Einstein and later Hawking remain confused and unable to explain why the big bang happened when it did and how somehow nothing spontaneously produced everything. Einstein, Hawking and their drones never understood the fractal nature of the field and that he field is composed of diametrically opposed, conjugate lines of force which were collapsed together BECAUSE of the Pythagorean discovery of Incommensurability in the Golden Ratio found everywhere in nature. The golden ratio difference in holography is the demonstrable action which collapsed magnetic, elastic, infinite lines of force to collapse our local region of the field. The collapse produced a massive, quick, blowout, elastic line entangling EXPLOSION of energy. The product was tangled, opposing lines tied together, against their natural resting state, i.e.

ATOMS, seemingly negative and positive & neutral in random ratios. How did these geniuses miss it? There were many contributing factors to that question. Which model is correct? Einstein's particle based "time-space" fabric which eliminates the reality of electricity as explained by the field experts who worked with it successfully. Maxwell, Faraday, Tesla, Heaviside, Steinmetz and more were closing in on this understanding of the field presented in this work. Compared to these scientists Einstein was a dilettante. Einstein contradicted them all and failed to account for the dielectric lines of force and how they joined with magnetic lines produce ATOMICS!

To verify my analysis of Einstein read his own works which tell you the same things that I have in…

1. The General and Special Theory of Relativity
2. The Problem of Space, Ether, And the Field In Physics
3. A Brief History of Time; Stephen Hawking

A final clarifying note, yes electrons exist as do neutrons and protons and therefore chemistry equations involving them same ae valid. The question isn't whether they exist, the question is what are these "particles" really made of? They are tangled lines of force & that is why action on a particle in one location affects another, miles away, instantly because both are tethered as part of the same elastic line of force!

CHAPTER 5

Post Isaac Newton's Gravity Mystery, A Thing He Couldn't Explain, as Gravity and Anti-Gravity Are ELECTRICAL!

Sir Isaac Newton, is a go to scientist in modern physicists, he is the grandfather of modern physics, his three laws of motion still hold true today for the larger part. Isaac Newton is perhaps most famous among non-scientific people for his observations and work of experimentation with gravity. Gravity fascinated Newton. The main body of work in which Newton codified the three laws of motion and contained the majority of his gravitational measurements and observations was a work which he published in July 5 1687. The work is **Philosephea Naturalis Principia Mathematica**. Here Newton outlined three important laws.

1. The law of Inertia: A body in motion tends to stay in motion and a body at rest remains at rest unless acted upon by an external force.
2. Force is equal to mass (a body) times acceleration, or F = ma (Force is required to cause motion)
3. For every action there is an equal and opposite reaction.

The above is the work which Sir Isaac Newton is best known for, but he wrote more and clarified more on the subject of gravity in his work entitled *Four Letters From Sir Isaac Newton to Dr. Bentley. Containing Some Arguments in Proof of a Deity* (1692). You won't find your modern physicists quoting anything from these letters because any quotes from this work will destroy their false narrative push which implies that Sir Isaac Newton and his laws are the solid basis upon which Einstein drew on for his ridiculous special theory of relativity. In the theory of relativity Einstein revives the premises of Democritus and Leucippus concerning the void as being a principal, a thing. Einstein implies that "space" which is not a principal, is not a thing, has curvature and this curvature is what is causing gravity, a complete oversimplification which fails to account for sustained orbits instead of everything falling in together in short order. Isaac Newton and his laws are implicated into Einstein's ridiculous theory to sell it to the clueless public and sycophant academicians. Sir Isaac Newton never claimed anything close to what Einstein implied about curved space, or a "time-space continuum" or anything about quanta of particles affecting one another at a distance by this same weird idea of "curved space". Newton was not so pretentious as to reify a posterior attribute, a non-principal privation, which is how space is to be understood. As Ken Wheeler states, space has no properties, it isn't anything but is much like a shadow which is not a thing, not a principal but is only a privation of light as Aristotle clarifies in the *Physics*.

Sir Isaac Newton was a Natural Philosopher and a truth seeker, unlike those in Physics today who gush over him and misrepresent and mischaracterize what Isaac Newton was stating regarding gravity. The modern physicist teaches us that Newton was a master at explaining and denoting what gravity is, that he "figured it out", revealed its hidden truths. This is not true; this is not what Newton stated about his own work at all. What newton was doing was measuring and observing gravity so as to best make use of it, he was not in any way explaining *what it is* and *how it works*, just that it does work and is useful and measurable and quantifiable for calculation. If you are going to believe anyone on the subject of gravity believe Sir Isaac Newtown himself... (In the Letters Newton states)

"I have not yet been able to discover the reason for these properties of gravity from phenomenon and I do not Fein a hypothesis known. For whatever is not deduced from the phenomenon must be called a hypothesis, whether metaphysical or physical, or based upon occult qualities or mechanical have no place in experimental philosophy. In this philosophy a particular propositions are inferred from the phenomenon and afterward rendered generally by induction."

But wait Newton gets more specific...

"it is inconceivable that inanimate matter should, without the mediation of something else not material operate upon and effect other matter without mutual contact...gravity should be innate and essential to matter that one body may act upon another at a distance through a vacuum... without the mediation of anything else through which the action and the course may be conveyed from one to the other, it is to me so great an absurdity that I believe no man who has, in philosophical matters, a competent faculty can fall for it... Gravity then must be caused by an agent acting constantly according to certain laws, but whether this agent be material or immaterial, I have left to the consideration of my readers.

These quotes were brought to our attention through the research of Mr. Ken Wheeler, Metaphysician and Natural Philosopher.

"Agent", what is this agent which Newton is referring to? Well it is not "curved space" but it is the very thing which Aristotle was referring to as the substratum which is the CAUSE of the four elements. Like Einstein, gravity seemed like spooky action at a distance. Spooky action at a distance, what a profoundly unscientific thing to say! Isaac Newton lived prior to the electrical age of discovery, experimentation and invention. His laws and understanding must be cast in this light! If Newton had been contemporary with Tesla he would have found how his distant bodies in a vacuum were talking.

Online, on the internet is a video, one of many, which supplies a good visual demonstration of fact that gravity is electrical in nature. This is important to distinguish as Sir Isaac Newton was conducting his observations and research just prior to the electrical age of discovery and experimentation and therefore accounts for his failing to make the connection, elsewise I am confident that he would have! The video which demonstrates the nature of gravity and how we manipulate it by means of dielectric and magnetic lines of force is clear for all to see. The author of this video is listed as this...

The video is entitled...
Copper's Surprising Reaction to Strong Magnets | Force Field Motion Dampening. This video was published online at...

NightHawkInLight
Published on Jan 26, 2018
The bio reads...

In this video I experiment with Lenz's Law And Faraday's Law of Induction to generate electricity and magnetic force fields in copper. Check out my sponsor Brilliant for a really fun way to learn! https://www.Brilliant.org/NightHawk To read more about Lenz & Faraday's Laws see the following links: https://en.wikipedia.org/wiki/Lenz%27... https://en.wikipedia.org/wiki/Faraday... Thanks to all my Patrons for helping me make these videos! A special thanks to my top Patrons: Syniurge, Matthew Leitzke, TheBackyardScientist, Enzo Breda Lee, John Johnson, & Thibaud Peverelli! https://www.patreon.com/NightHawkProj... Thanks for watching! SHOW MORE

YouTube video, found at the link... https://www.youtube.com/watch?v=sENgdSF8ppA&t=360s as of July 3, 2019 but of course will eventually be taken offline, thus the date stamp.

For those who have not watched the video understand that in the video our host sets up two copper tiles horizontally on top of each other. The tiles are solid coper squares resting on a wooden structure. Circular magnets are set on top of the copper plates; a second small cube shaped magnet is set under the plate and is manipulated by hand underneath the plates. As a result, against gravity, the circle magnets rise up and levitate about a half inch above the copper plates. This is not the action of particles though the host thinks so and says so. He believes that electron streams are being moved through two inches of solid coper by the lower magnet as if the copper wasn't even there. Of course this misconception is based on particle physics; particle physics tells us that particles moving are causing this action. This is incorrect and actually disproven further in the video.

Further in the video our host stands the same two copper plates up vertically and he sets up the cube magnet more than a foot away from the plates, he then holds a second small magnet maybe six inches away from the copper plates on the opposite side and begins to spin said magnet by hand. Instantly and at a distance far outside of the magnetic reach and field of the cube magnet, it begins to spin in response to the hand held magnet's movement. This is impossible by means of particle physics. What we are watching is electrical; we have the atmosphere of the dielectric fields surrounding the copper that is why copper is used in motors and generators. Along with the copper we have the magnetic lines bending around the ferrous magnets. These are the components of electricity especially when set in motion by force of action. This too is demonstrated in the same video when the round magnets are later dropped through a tube surrounded by copper windings and those windings are tied to the leads of an L.E.D. When this happens the L.E.D. lights up; thus visually demonstrating the electrical action and interplay of dielectric and magnetic lines of force.

Image in if Sir Isaac Newton had witnessed the same experiments and demonstrations of the field in action. Imagine his notes on observing anti-gravity action being produced with the components of electricity. This would have answered Newton's question of how distant bodies talk to one another to affect them to gravitate to one another.

CHAPTER 6

TESLA the Misunderstood Bridge

Tesla is the missing link in our understanding of the Field. He is the culmination of Faraday, Maxwell, Franklin, Heaviside's, Galvani's, Aldine's, Volta's and many others research into the Aether and the electrical phenomenon. His conclusions yielded the electrical products and realities that we take advantage of today. Tesla understood the Aether, that it was charged at the entanglement points with various forms Electromagnetic expressions of radiation. He, like Faraday was coming to an understanding of the fibrous or cellular structure within the Aether and that this was the basis for the different size and energy level expressions of EMR. Tesla is the man who stood in the gap, who brought our race out of the dark ages of science. Tesla understood what his experiments meant and how they unified all of our understanding of time, space, history, electricity, contact with other worlds and much more.

Tesla did not agree with Einstein on his understanding of space and time. Whether Eisnstein was well versed in the philosophy of the ancient Greeks or not I cannot say but Tesla surely was. Telsa knew the logic proofs of Descart, Plato, Aristotle, the Pre-Socratic philosophers, about atomism among the Greeks and all the logic proofs that had already been proven and disproven. As to where Einstein promoted that there is no "Aether" it is Einstein who turned around and ascribed the attributes and juxtaposed nature of the Aether which Tesla was experimenting with, to two principals which he called time and space. Einstein applied the equations used to understand electrodynamics to "time and space". Tesla instantly responded to this lark explaining that time is not anything but positional motion, as the Greek philosophers already broke down with logic, and that space is not a principal as it has no properties in and of itself. Space is literally nothing but a giant shadow. A shadow is not a tangible principal reality but is rather perceived as a "thing" only in so much as it seems to have presence as a absence, or privation, of light!

This is the answer, a simple clear and elegant and final answer to the false physics which spun off of the public's ignorance of philosophy and ancient science. Tesla undid quantum physics with his simple answer but everybody missed it, it went over everyone's head! Space is a giant shadow which hides or masks the giant grid like Field of Aether composed of magnetic lines and dielectric lines positioned orthogonally to one another so as to form a giant, endless grid. Within each grid, Tesla found a size range of electromagnetic lines which come in a range size. The range sizes turned out to be Radio, Microwave, Infrared, Visible light, Ultraviolet light and gamma in braided size configuration. He also believed there were Cosmic EMR rays. Tesla was right and his experimentations with success gave us the wireless transmissions we use today to even send signals through space which is supposed to be a empty vacuum. By the definitions of electricity supplied to us through Einsteinian based physics wireless transmission cannot work, yet it does! Thank you Mr. Tesla. Who should you believe in this debate as you probably are reading this information on a Tesla based wireless device?

Tesla, Nichola Tesla, was the premiere Electrical genius of the nineteenth and twentieth centuries or was he? This book is not a biography, but an introduction to Nichola Tesla beyond his being an "electrical genius". To say Nichola was an electrical genius sounds like a compliment and it is but in reality this title is a backhanded compliment as Tesla was this and so much more. This book begins with the assumption that the reader already knows much of Tesla's biographical story and also shares the majority opinion on Tesla, that he was an electrical genius. With this expectation stated I will jump right in and tell you that Tesla was an electrical genius but that electrical work was a means to an end, a higher end. Tesla's work in electricity has not been matched, however much of his other work gets ignored and glossed over, even covered up at the expense of categorizing him as an "electrical genius". You see Tesla invented many, many things. Many things were seen in his lab and were poorly, to completely, misunderstood by all the spectators. Tesla invented many other things apart from overtly electrical things but electrical generation was a necessary component part of those other inventions. Tesla was using electricity to do many other things. His main interest was broadcasting electric power to every person on earth, wirelessly, and for free. He wanted first and foremost to literally empower the people, to set them free physically from hard labor. He believed that with abundant electrical power the worlds machines could do the hard work and that people could then be liberated to realize their higher purposes by pursuing philosophy, science, art, philanthropy etc.

Almost no one will tell you about other things Tesla invented but did not ultimately receive the credit for. In case anyone tells you anything different be assured, Tesla invented the X-Ray machine. Tesla invented WIFI, wireless communication. Tesla invented infrared remote control. Tesla invented radiant rainbow light. Tesla invented EMR devices which the military uses as a death ray, heat ray, a burn ray, an EMR ray which is adjustable to make crowds feel uncomfortable all the way up to fried alive in their skin. Tesla invented Radio, radar, radio communication and power transmission, wireless transmission. Tesla may have invented ultra-violet sterilization systems as used on surgical equipment. He found a connection to gamma radiation and Marie Curie's radioactive elements. Tesla's work in electricity was a means to an end. Tesla invented the Tesla coil to fulfill all these ends. Tesla coils are at work in all of the above mentioned inventions. Tesla coils of different sizes and windings vibrate, move or oscillate at different speeds, rates/frequencies. This is the key that unlocks Tesla and what he was doing overall in the big picture. Tesla was using Tesla coils to their fullest range of potential on the full EMR spectrum. He was trying to find the boundaries of the electro-magnetic wave formations and their size and application limitations, both in large and small wave sizes.

Through Tesla primarily we learned that the largest EMR wave formations would be what we call Radio waves. Radio waves are a braid of large magnetic lines, braided with dielectric field lines. Radio waves are the largest ones that exist. Dielectric grid/field lines exist orthogonally/perpendicular to the giant magnetic lines of force found throughout the universe. These lines of force are a feature of an omnipresent field ubiquitous in the universe. This is the point of contradiction between Tesla and Heinrich Hertz. Hertz had theories of wireless transmission which were adopted by Einstein and most physicists which Tesla tried to verify but could not. In trying to test Hertz' theory of electron movement forming fields around them he found their theory failed in application. Their wireless transmissions were only successful over small distances thus making them theorize that electrical fields were formed around electrons being projected. Tesla proved this idea false when he made wireless broadcasts that traveled miles and later even on through the vacuum of space. This should and would be impossible if Einstein, his physics sycophants and Hertz were correct about the nature of electrons and electrical fields.

It is EMR which proves that Einstein's model of reality is a farce, a lie, a monstrosity of misunderstanding and that a net, a grid, a field, like a giant Cartesian coordinate plane is real, extant and ubiquitous in the universe. This field, inside its grid lines contain, smaller fields, recursive versions of the larger field going downward asymptotically without end. A field of fields, grids of energy within grids of energy all different sizes existing juxtaposed and contemporaneously with one another without bottom or end, a truly Sisyphean field of fields. Our reality, our material universe is constructed out of lines of force which range in size found in Electro -magnetic radiation in all its known manifestations of our realm. Radio waves, are the largest EMR. "Microwaves" are the smaller version of Radio waves and are thus called "micro" even though they are the second largest wave form. Down from Radio/Microwave EMR is Infra-red "light". Then "Visible Light" follows in EMR wave size. After visible light comes "Ultra-violet" "light" EMR, a.k.a. UV light. After UV light comes X-Ray EMR, then finally there is Gamma Radiation EMR. The last three can be quite dangerous to life, cells and DNA if exposed to them. They are smaller wave forms that are high energy and they ram into DNA and reorder or break the lattice all together.

Tesla found all these EMR expressions by tuning his Tesla coils, it is why he was making Tesla coils, it is why people could walk into his lab hearing a Tesla coil humming and seeing every color of light found on the rainbow lighting up his lab WITHOUT any light bulb or tube source. People didn't understand this then and still don't today. These sound only like hyperbole, exaggerated anecdotes, they are not. They are what I have been describing. The so called "death rays" truly are EMR wireless transmissions/projections off of a Tesla coil exited to oscillate at extremely high frequencies far above the visible light oscillations. If built correctly it is merely a volume control button which can cause a properly wound Tesla coil to radiate visible light in various colors to projecting and aiming the same device to project X-rays or even Gamma rays. It is just this simple.

People today who are mystically fascinated with Tesla build Tesla coils of their own. What do they do this for? For sparks, for a lightning show. They don't even know what Tesla coils are for other than pretty sparks. Tesla coils are for one general function, to vibrate a secondary coil or receiver of primary broadcast signal. The wave form and size produced by a given Tesla coil depends on the number of windings and voltage applied and produced and possible intermittent interruptions creating oscillations. Oscillations simply refer to rate or speed at which a coil is vibrating at. Higher frequency, meaning faster and more often vibrations, produce more energetic, smaller waves. Frequency literally means how often something happens over time, like how frequently do you exhale in hour on average? The answer; six hundred times, for example. Different frequencies produce different size wave formations, for TRANSMISSION, which means to send over a distance wirelessly. People today talk a lot about 5G WIFI transmissions not usually understanding that these transmissions are microwave sizes radio signals broadcasting the full contents of the internet. Television signals are carried on radio waves which are oscillated and created by means of Tesla coils. Now you know.

Perhaps it is long overdue that some of us learn to build and tune Tesla coils for variations of speed/rates or oscillations. Possession of such a coil or set of coils could render one person one of the most powerful beings in the world. Tesla's work on the Field, research into the Field, the nature and structure of the Field is the breakthrough to understanding all the mysteries of matter, energy, time, space travel, anti-gravity propulsion, time movement, the Ethereal realm and all entities of consciousness not to mention the essence of life and death itself.

CHAPTER 7

The E.M. Spectrum Wave, was the "Big Bang", a Traveling Wave of Intelligence, Energy and Consciousness Passing Through our Static Measurable Portion of the Field Imprinting a Reflection of Itself in Matter

Consciousness, it needs to be defined. Consciousness usually is defined by the discipline of psychology. Psychology defines consciousness as a product of higher intelligence and a byproduct formed out of necessity by evolutionary processes. Evolution formed consciousness as a greater tool set to increase the chances of the species in question of being able to survive, by means of being better able to gather food and protect said food. An example in nature of consciousness evolving is when birds gather nuts to save them for eating in winter or a later time. Consciousness begins to manifest as an awareness of others outside of yourself and recognizing that others are thinking thoughts and planning actions as well. The birds that hide nuts at times can be observed exhibiting just such consciousness. Birds that are hiding nuts sometimes notice that other birds are watching them do this activity. The birds who observe the spy birds also show intelligence in assuming that the other bird is spying so that it can and will steal its hidden nuts as soon as the working bird leaves. The bird who observes that a spy bird is watching his activity, will depart, then circle back once his adversary leaves, and then will go back to the hidden nut, remove it, and then hide it elsewhere. Such intelligence and awareness of others and self in relation to them helps the birds better survive and always be fed.

Intelligence also manifests in our example birds when birds are observed using tools like a stick to aid them in digging for ants. Birds have been observed using a piece of wire to dig as a tool and then bending said wire, actually modifying the tool. Again, this example of intelligence in living animals clarifies the point that such evolutionary manifestations are for the primary purpose of survival. It could be explained that this is why consciousness manifested so prominently in humans.

Consciousness the Untold Story

While *consciousness* is often equated as an evolutionary innovation in biology, which only manifested in animal life relatively recently, this assessment is *not* accurate. What you have never been told is that consciousness in fact did not evolve with the creatures and plants as they evolved into more complex forms. In the very earliest and simplest life forms, Cyanobacteria, was found already impossibly complex, environmentally "triggered" DNA chains of a complexity order still beyond human capability to replicate or even accurately map with the best of all of our computers. Realize that our computers, which are set onto the task of calculating all the variables of Cyanobacteria, with all that their switches, and what all they can manifest when triggered, by various environments, cannot as of yet, equal (or match) the level of coding involved in DNA of a "simple" bacteria. Our computer programming represents the pinnacle and maximum demonstration of our deliberate conscious application of intelligence through logic to produce an elaborate code. The code of the simple bacteria, as such, DID NOT EVOLVE with the creature. The DNA code is programing at the molecular level which includes the possibility to manifest creatures, still as of yet unknown and unseen. Erroneously creature manifestations not yet manifested are designated NOT real, but not real is only a relationship as it applies to time. When we say that something is "NOT REAL" all that really means is that it is not real YET! This implies that literally every conceivable creature can exist, either here in a material "Living" manifestation or in the light spectrum where not possible in the matter. Consider what this implies, obvious intelligence and consciousness has been present and evident since the first manifest flora and fauna! Odd but true. Weigh with logic what this means, the very highest order of applied, deliberate, directed, conscious human intellect is showing up as still lesser in comparison to that found evident, real and present in a bacteria found on earth two billion years in the past! Consciousness is a great big pair of shoes that creatures on earth have been growing into gradually over time. We have evolved to a point where we demonstrate consciousness and intelligence but it has been here even before the Cyanobacteria!

Biologists have attempted to mechanically reconstruct the evolutionary process as it may have unfolded in matter. They do this with zero real understanding or consideration of the E.M. Spectrum as a whole acting on matter. They know nothing about the Field and how it organizes. Biologists inform us that among the simplest and earliest life forms on earth were the amoeba, a single celled life form. As best as anyone can figure out biologically, amino acids were rich in the early ocean floor around the warm water vents. These assembled the early chains in the form of RNA. The RNA is said to have risen to the top of the oceans and on hot days would be pulled up in an evaporative process. Exposure to the sun changed the chemical composition of the RNA to DNA by the conversion of Uracil to Thiamine. The new DNA ended up back in the oceans through rainfall. The DNA occasionally came to settle in lipid rich pockets and puddles which would eventually house the DNA as cell walls. This progression explains how life would have to have originally manifested and formed the DNA. The origins of this process are estimated to go back to about two billion years ago. This is a fascinating analysis and of necessity has to have some accuracy to it.

Assuming everything played out just as described, a DNA chain in a lipid cannot organize itself to a predetermined end. A pre-determined product, or set of products, was in mind when the DNA nucleotides were organized by intelligence. Someone intelligent noticed that nucleotides can bond on spines to form a binary helix. Someone noticed that nucleotides could be organized to produce a product when interacted with certain EMR. Someone or some intelligences engineered a program out of naturally occurring RNA strands into product producing and self-replicating DNA. DNA did not and cannot make the quantum leap by itself to organize into cell based life and never have done so. DNA cannot self-direct to assemble in a configuration that will result in a highly organized plant or animal which wants to survive in its lifetime and is capable of adapting to different environments. The best lab tests reveal the same thing. DNA can only naturally occur without intelligent manipulation up to a basic point and non or that process results in any product with any coherency.

Evolution is the science of measuring and observing a gradual series of biological changes (IMPROVEMENTS) in creatures which manifest ever increasing complexities in compositions and capabilities in order to better survive in different environments over time. Different environments trigger the DNA to bring out manifestations of *IMPROVEMENT* in creatures. DNA, with how complex it is, and what it is, and what it moves to do, and function like, goes back nearly as far in time as two billion years. Evolution has been slow, but DNA complexity was a rapid formation in matter. DNA in great complexity, reaching deep back into prehistory, cannot be said to have been formed over long periods of time the way that organisms did. This is bizarre. DNA is directed, directional, seemingly engineered and constructed to manifest improvements to the survival capabilities of the larger organism. These are the reasons that Christian and creation "Scientists" claim that life was created by "God". Christians fail to acknowledge however that the scientific analysis of the code of life, the DNA, in no way coordinates with the Biblical Christian narrative of how their God formed and created all living things on earth and it definitely does not coordinate with the given timeline. The early complexity of the DNA and what it does and manifests and has pre-built into it to manifest in a progression of complexity, remains unaddressed by biology.

The truth is no one can address the origins of consciousness with studies of *biology* or *psychology* because the story of consciousness begins *before* biology existed. Biology is a byproduct, an effect of

Intelligence and consciousness (In the form of the E.M. Spectrum) impressing itself in the slower medium of matter! Intelligence/consciousness, presented a manifestation of itself in our portion of the E.M. Field. Remember the big bang, caused two distinct things to come into existence 13.8 Billion years ago! The Big Bang brought the E.M. Spectrum UP into our discernable/detectable portion of the E.M. field from the smaller higher energy portions internal to the Field. The second, different product the Big Bang generated was matter, the material, the so called atomic particles. A simple definition of atomic particles can be thought of as no more complicated than tangled lines of force! The field, the portion of it which we can recognize in all of its size variations, consists of microscopic/"subatomic sized" to building sized radio wave sized magnetic/dielectric lines. The reason humans have had so much difficulty discovering the origins of consciousness is simply because consciousness had no material beginning! Consciousness is not a created thing with a beginning! It's not a product of creation, evolution nor engineering. Consciousness is a natural phenomenon eternal in status, a product of the field which has no single origin nor terminal. Consciousness is an effect on static lines of force in the field rolling, compounding and gathering in complexity. Consciousness is the imbalance of the field which makes it synonymous with motion. It is in motion because of its conjugate positioning in relation to itself, to its counter lines of force. Don't let this statement be twisted to interpret a god into this equation. Deferring to a god at this point is only intellectual laziness. The story of consciousness, kind of "begins" when in our detectable portion of the field's, at rest (static state) lines were disturbed by the big bang. As covered in an earlier chapter, modern physicists tell us to believe that nothing conceived

145

everything, but as Aristotle makes clear that <u>nothing comes from nothing</u> and begets nothing. Ex Nehilo, Nehil Fit. Of course the Big Bang is not a case of everything, time, matter, space and energy coming into existence from nothing! This idea is plainly and simply asinine! This belief is also the belief of the religious.

So far we have been discussing the field, with the lines of force, magnetic and dielectric in an x, y, z grid. The field, the Electro-magnetic Field, is not the same thing as the Electro-Magnetic *Spectrum*. Even if people blur the two things together they are not the same thing. The E.M. Spectrum looks something like the following graphic...

The size of the EM spectrum

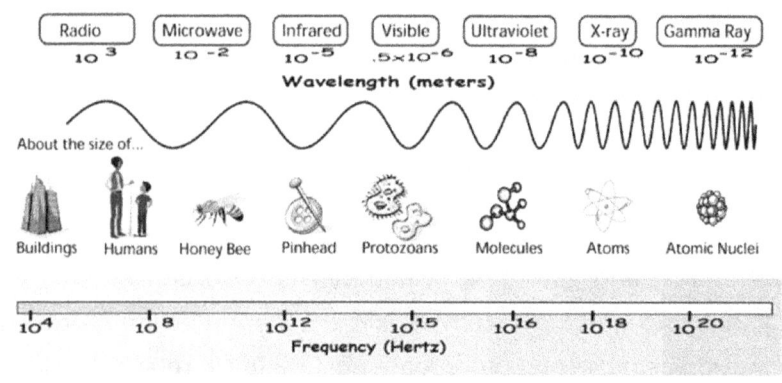

Chapter 6, fig. 1

The E.M. Spectrum (of Radiation) as seen above shows, or describes, the relative wave sizes but you need to see further that the waves of all these different sizes are the interplay of electric and magnetic lines of force acting together as seen in the next image...

Electromagnetic Wave

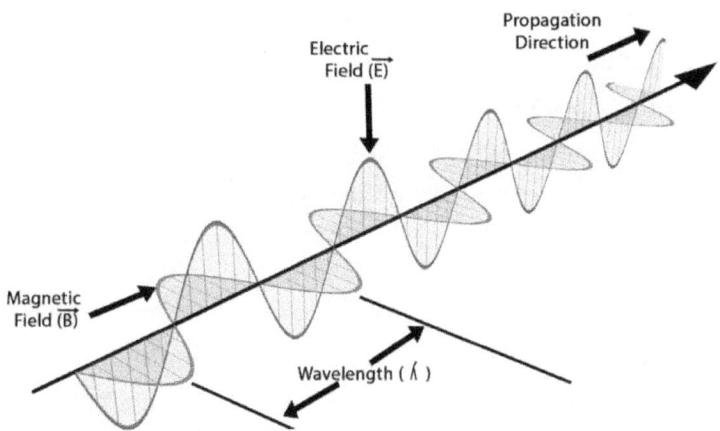

Chapter 6, fig. 2, All sizes of E.M. Radiation can technically be classified as "light" which is confusing as we tend to think of "light" as what we see and what helps us see. Visible light is just a small portion of E.M. Radiation types which we happen to be able to "see".

The wave form which you are looking at comes in all the different sizes as seen in the previous graphic. The interaction in all these sizes was vibrated into interplay, into interaction by the big bang. The Big Bang is something that happened to the, at rest, E.M. Field or grid. The spectrum came to our detectable, discernable part, of the field, from a much smaller part of the Field. As light studies have revealed through holograms, light images recorded on glass can capture the image of anybody or anything on the whole plate. Holograms demonstrate that the ever descending E.M. spectrum

147

of lines continues downward into infinity, like ever smaller dominos in a downward descending line, into what the Pythagoreans called "Incommensurability". Hologram plates can have a green laser shone through them to reveal an image. If such a plate was cut in half and a laser is shone through that half, the image which it reveals is the whole image again in a diminutive form. When you take that half plate and then break it in half again and shine a green laser through it, guess what? Yes the whole image and all of its image information appears again in a scaled down version. This phenomenon goes on repeating downward in scale ad infinitum. You are seeing the grid, or endless grids of the Field.

We live in (our material universe) a certain portion of the E.M. Field; the truth is that the Field has neither size limitations nor boundaries. There was a time, about 13.8 Billion years ago when the travelling wave of motion reached our DISCERNABLE portion of the field from the infinitely smaller (higher energy) portions of the field. The traveling wave that reached our knowable portion of the field, literally, never had a beginning nor end nor will it terminate anywhere at any time! This is not something people can process very well, it is contrary to our need to find an origin and beginning, there never was one! Take a minute with that thought. The day that our portion of the field was acted upon by the traveling wave of motion was moved into motion by an imbalance of size in the magnetic and dielectric paired lines! Here is the mind twisting portion of this whole analysis, the E.M. Spectrum, the wave, or waves, in our field are the basis and material of consciousness itself/intelligence itself! The intelligence has always been there, compounding, overlapping, always rippling through the field like ripples in a lake or pond when a rock is thrown into it! This information about the E.M. Spectrum prepares us to start understanding what DNA even is and why it is imbedded with very complex, directed, intelligent and consciously predesignated, and goal oriented with coded information.

While DNA and matter decay, break down, and the things which they manifest in the realm of the material "die", amorphous intelligences in the E.M. Spectrum formed the DNA, remain as in an underlining substratum. The E.M. Spectrum is the pattern substratum which shares a substratum with atoms in all their manifest states of matter. The E.M. Spectrum impacts, imprints and shapes the Material manifestations. This happens much, as light digital storage technology does imprints information digitally in a solid medium, similar to how a CD/DVD burner receives light information.

Correspondingly and metaphorically he DVD Burner corresponds to the DNA material code.

The light digital information encoding fed to the Burner corresponds to the EM Spectrum.

The imprinted with light impressions disc, the DVD corresponds to the medium of the material universe, the atomic.

Much can be learned from this metaphor, that the lives we live or play out are pre-written in the E.M. Spectrum with little variance. This lends some credence to destiny and fatalism. This can very well account for manifest examples of genius from seemingly out of nowhere. We, the living creatures are formed after coded light impressions to manifest a movie (primarily in the form of our DNA), which unfolds the story of our lives. The DVD will become separated from the DVD burner/player; this is equivalent to what we call death. The DVD burner is writing the movie onto the unwritten DVD, the movie also can be played out on the same DVD burner but when the movie is over it is separated from the DVD burner/player, the termination of that story has happened and a new movie takes its place. Anything with DNA that has lived was impression burned and recorded with all of their uniqueness onto the E.M. Spectrum for all time to push forward, in one direction through yet undisturbed parts of the boundless E.M. Field. Long after the DVD burner is gone (the DNA and resultant lifespan) the individual DVD will exist with its eternally patterned impressions. The traveling light information that was impressed in DNA continues to travel through and away from the old material manifestation. Consciousness travels, what portion of it which was individually yours will be taken with the larger moving wave for all time. Consciousness is definitely perpetual but none of this should be misunderstood by the religious to imply that you or any creature has a "soul". You, and your consciousness and your lifespan are part of the whole of consciousness; you are only a separate partitioned bit of consciousness while you live your life out as an individual separated "cell" in your mater (material body). When consciousness separates at the time when your individual body dies, then your isolated bit of consciousness is poured back into the sea of consciousness which travels through the E.M. Field and the conscious you knew will go where it goes in a wave. You have made a recorded impression of all that you are in matter that was always there. Matter (because it is slower in propagation) traps consciousness in its matrix, temporarily, in isolation cells (lives). Matter acts as a retardation of consciousness, slowing it down, it takes years for consciousness to fully manifest in matter, material minds, and rediscover its orientation.

Light, or Electromagnetic Radiation is poorly understood compared to what still needs to be. The facts about the E.M. Spectrum need to be learned, all about all its properties. Consider, if you will how all that you see on TV, or that you access over the wireless internet web, are broadcast to us by means of *microwaves*. Microwaves, <u>contrary to their name</u>, are actually a very larger variety of E.M. waves. Microwaves should actually be renamed "MACRO-waves" as Microwaves are only small compared to radio waves, the largest of the E.M. radiation wave forms (detectable by us). Most of our highest technological digital storage information and transmission comes by means of these larger E.M. wave forms. Think about it, the sum total of human intellect which is transferred and broadcast is done so primarily in the larger wave forms. How much more life, intellect, imprinted worlds of NON-Material reality, can easily resides in the ultra-violet EM Radiation, the X-ray radiation, and the Atomic nuclei sized gamma E.M. radiation and the yet smaller Cosmic E.M. radiation? I'll tell you, the magnitude is literally beyond all of humanity's ability to imagine and conceive of, even after millions of generations. You just cannot wrap our sum total conscious efforts and intelligence around a fraction of its reality patterning and imprinting abilities! I'm sorry, that may sound like pure hyperbole, but it isn't.

Now focusing on DNA we must realize that DNA is not the E.M. Spectrum itself, it is not composed of it. DNA is consciousness's hand print, or reflection, impressed in wet sand which leaves a mirror impression. DNA is material, constructed with amino acids arranged in proteins, composed of molecules and atoms. We have to understand what material matter actually is. What are atoms? What the secret of the particles is as found in the study of "Particle Physics"? This was dealt with in chapter three. Particles are not particles in a vacuum as Democritus, Leucippus, Newton and Einstein misunderstood. Aristotle more than two thousand years ago demolished the misconceptions of the existence of the void of the atomists. Atomist logic was dismantled and disproven by Aristotle over two thousand years ago. The day of the Big Bang is the day that all the so called particles came into existence. Particles are a manifestation apart and independent and separate from the E.M. Spectrum which was a manifestation of <u>untangled</u> lines of force set in motion. The particles are <u>finite</u>, and <u>quantifiable</u> in number, this much is true but what they are composed of is <u>infinite</u>. The fact that particles are finite and quantifiable is what Democritus, Leucippus, Einstein and Newton loved about them and the subsequent theory they use to explain all reality with. An Aether or endless Field gave such men anxiety, their minds cannot consider "real" what has no finite volume, yet they believe in infinite "space" with no problem. Finite particles are part of the infinite Field, deal with it! What particles are, are tangled magnetic and dielectric lines, a natural result of massive heated expansive explosion! Some tangled lines are lines of the same size, some of different sizes. The different sized tangled lines are what we perceive as "ions", as in positively and negatively charged atoms.

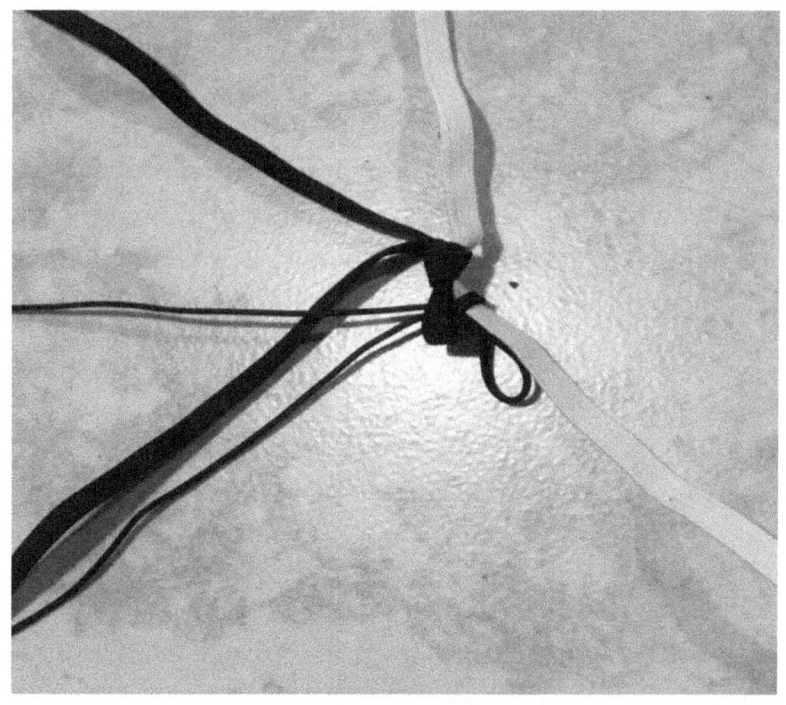

Different Sized Lines Tangled would favor the type of Line which was smaller and higher in energy.

Matter and atoms, the material in all of its states, are things apart from the E.M. Spectrum. The E.M. Spectrum is what is encoding life and form into its mirror manifestation into and onto matter! Matter is inferior to the E.M. Spectrum and is finite in number and held form. DNA is encoded by the E.M. Spectrum, statically attracted to it to form, DNA with all of its encoded information, with all of its ranges of possibility. DNA is essentially a glove covering the unseen hand, the shape of that unseen hand (E.M. Radiation). DNA manifested early on in the encoding of the amoeba, complex bacteria and all animals all the way up to mankind. Mankind to this day cannot encode a binary code which self-replicates and switches on at as many gates for as wide of a field of variations as the DNA of a "simple bacteria". Even with the highest and most predominant manifestations of consciousness and intelligence as found manifesting in mankind we have not achieved the same level of intellect, conscious intellect, reflected in our encoded programs as that simply reflected in the simplest of DNA! The intelligence and consciousness of the EM Spectrum is still much smarter than all of the rest of us and best of us are intellectually and consciously combined.

Observations of the Fossil Record, What Dinosaurs and Other Ancient Creatures Reveal

Even an amateur who studied the fossil record which shows the evolutionary progression of life on earth cannot miss just how seemingly engineered and experimentally deliberate and end goal oriented the manifested life forms appear to be.

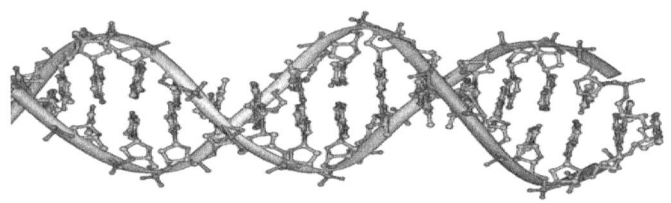

This is what DNA looks like, why follows

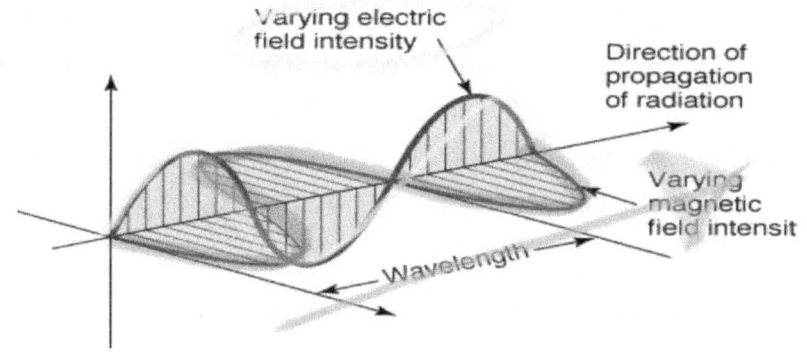

DNA, it's mirror image

Carefully observe how that evolution is directional. Observe the animal life forms how they are first simple in form and gradually transition to more complex forms. It is as if *someone* is working out the design which will eventually manifest intelligent beings with conscious works. Someone, plural, and conscious had to be directing. Randomness will not direct to an improving product. Who is directing and organizing the DNA beyond its natural limitation to organize? Could this be the God of the Bible, or another religion, space aliens? Who? The answer is complex and is partially <u>the field itself</u>! The field is NOT evolving, but the field is compounding and exponentially learning. The field has products of compounding intellect within it. Within the Ethereal Field are life forms, intelligent and lesser so, plants, mountains and amorphous beings. These encountered the slow material realm, the portion of the Field forming atoms, the slow realm and took it upon themselves to imprint on the material realm what they saw in Holographic reality. These are neither moral nor amoral, neither good nor evil as we understand those subjects. These are the older children of the Field, they are a compounded product of EMR intelligence and consciousness. These encountered us in the holographic realm, that is where we are rooted and are from, that is our home. We are the intelligent, but newer, immature in comparison, young. They made the DNA encode our images so as to throw us into its matrix to process us once, twice or numerous times in earthly lifespans to mature us.

In the past the elder children of the Field, in the Field, experimented for billions of years with the DNA code in various environments to engineer housing in bodies adequate for beings of light to cycle through a lifetime in. Primitive and superstitious humans have called them "gods" and depicted them in a multitude of forms. They have no native form but they made utility material bodies for themselves as well. They did this to more easily interact with the materials of the atomic realm. The DNA was used to grow vessel, avatar bodies for themselves. They used biology and nucleotides because through manipulation of EMR they could bond and encode digitally a program which could be used as a software that would essentially 3-D print them a machine body to interact with on in this realm. Some misguided people and governments now call these beings "aliens" from outer space. They are not "alien" to earth and they are not "space men". Once they had the bodies they had the tools to mine the materials to make the craft they move around in. This is information the governments don't even know.

The amorphous beings are from the Field. They are the compounded product of intellect and consciousness and they are dominant in the realm of the holographic Field. They are mostly interested in our improvement and maturation but some are more amoral than others and may be younger and more immature among their ranks. The technical term in western society for the process of maturation and the need for humans to begin the process is usually "Messianic". The growing of the godliness within the man/woman. In the east the communication is also through a messianic enlightenment tool called Buddhism and the process is called "enlightenment". The names are just names, the process is to unify with the attributes of holiness and being better in general of "god". Religion has been the vehicle of primitive relaying of this process in primitive terms. The point of it all is to eventually return a matured, cleansed and purified and improved intelligent hologram teacher to the EMR realm. The realm isn't "heaven" as "heaven" is being produced by processing immature creatures to maturity so all can be in heaven together.

The field and the electromagnetic radiation is complete and already containing all that has happened and all that will happen, static, with all possible variations of outcomes! The wild card in this scenario is here, in the material realm, here people will purify if it takes a hundred lifetimes or just one. This understanding of physical reality returns us to our great ancient philosophers. Parmenides told us that the universe is not what it seems to be to our perceptions, that the universe is static, unchanging with its history pre-written and the manifestation of what we think we perceive is only us playing out what we always were going to play out. In other words nothing is gained or lost, only the wave in the field is in motion, it is pushing through matter, it is upending it, causing new patterns that rise and fall just as a kaleidoscope makes and brakes new images. Matter is mirroring, as best it can, as fast as it can what is already in the Spectrum of radiation. This is evolution and the life cycle; they are part of the field! Think about this, time is therefore the push through the field, it is one directional. Time, evolution and consciousness are clearly linked.

Yes DNA is complex beyond the ability of human intellect and consciousness to currently produce, yet nobody built it exactly. The wise engineering amorphous children of the field manipulated the DNA to mirror the pattern manifestations of the holographic field. DNA did not evolve, it appeared whole and complex long ago, already encoding all the possibilities of life we see now and in the fossil record. DNA is electrostatic, electrostatics have everything to do with the formation of the double helix of the DNA strand. Just as is the structure of all electromagnetic phenomenon along the E.M. The manifestation of the DNA we see materialize is very much to be compared to a footprint left in the sand. When we see the footprint we should not be in awe or resort to religious/superstitious speculation that God build the complex form out of the sandy material. That is insane; the print is obviously produced by a human foot of the exact shape of which the print is a negative of the original foot! The DNA corresponds to the footprint; it is the impression (imprint) on the medium of the sand but is cast of a real foot. The real foot is not made of sand at all. In the case of the DNA it is the imprint in a material medium of a real electromagnetic coaxial (Thank you Ken Wheeler) structure, probably of the ultra-violet range, as this is the most common E.M. radiation available in nature on earth in a moisture evaporated cycle. So who created DNA? DNA is not a creation it is an imprint left by something not created at all! E.M. radiation in all its forms has always been traveling through the incommensurate field. That being said, there was a bridge to the action occurring.

Death and dying involves the departing of the E.M. Radiation, dying is the releasing of longer nucleotide chains which causes eventual termination/death by shortening of the chains used for replication. This process is called the shortening of the telomeres. Very much the way a foot is set down in moist sand, evenly pressed in and then the heel is lifted out, then the whole foot proceeds to be rolled out by the toe area, so it is with E.M. radiation and nucleotides. As it is with sand and a foot so it goes with E.M. contact in polar charged nucleotides during the initial gathering for formation. The charge in the DNA similarly rotates through the medium of a nucleotide soup and then rolls out. This phenomenon results in what we consider aging in the DNA, the degrading and shortening of the telomeres. Amazing! Now follow the train of logic. The DNA strand and code manifests actual living creatures and plants. DNA materially is matching and being built on an electric spine or grid, what is manifest materially is a mirror of life in the light or E.M. radiation Spectrum and REALM! All the plants and animals and non-animal forming expressions of light or E.M. radiation is real and mirroring itself as best it can in the retarding and slowing medium of matter, proteins. Do you see? The light frame which gradually departs from the imprint in material amino molecules is not destroyed, nor ended even when the sand cast falls in after the foot departs. In other words, death and life are nothing of what we have previously understood. Metaphorically, when life begins and forms a the DNA level, material DNA is like one half of a Velcro strap being pushed together with the other half of the strap, the electromagnetic skeletal frame so to speak. When we die, or rather as we die the Velcro sides are gradually being separated.

It is Aristotle who analyzed first with logic that evolution is clearly evident, as Empedocles observed it as well. The difference between Aristotle and Empedocles was that Aristotle took into consideration the environment and concluded that evolution is NOT *random processes* manifesting random outcomes. God cannot be credited with creating matter unless you are willing to say God is the E.M. Spectrum. This generates a new problem if you think that then you must realize that the E.M. Spectrum is a product of the Eternal E.M. Field going into a localized motion! In other words God would then become a creation/product of the larger, unmoved Field and its lines of force. Matter is a tangled mix of lines of force. What we call life and death for material animals and conscious humans is the same rise and fall which we call the life cycle. The wave carries all that was from the past to the future evolved creatures in the Matter (material) world. This wave and all in it are part of the *zeitgeist*. They all live and they all die but they are evolving (manifesting diversity) as well. Life evolves *to better survive* and it does this but this is a contradiction because life also survives and perpetuates by pushing through individual creatures who strive to survive efficiently, ONLY TO DIE anyway. "Life" and consciousness are connected. If consciousness is not eternal then logic contradicts itself in the logic puzzle of why should creatures who will die fight to survive, who will die anyway. Logic dictates that we seek equilibrium and a resting state. If all animal life is destined to die why doesn't it just lie down and die? Why should animal life strive to survive? Why doesn't animal life stop eating and procreating immediately since doing so is a shorter path to the given inevitable state? Logic dictates that it is programmed into the DNA to do all it can to survive even if it means killing off its individual hosts. Death is illogical to conscious driven life formation, it is non-sequitur. Survival is paramount to DNA life and DNA life imbeds conscious intelligent coding which is not from the material, matter world. Why should the DNA code be manifesting, or turning on switches which will help creatures survive in different environments when no matter what they will die anyway? There is a why answer. DNA itself will cease, of course, when matter all comes to relative rest and largely untangles itself but the traveling wave is unbounded by fixed boundaries.

Consider what DNA has manifested anciently, the dinosaurs, what immenseness and powerful and what brainless creatures they were. In body structure dinosaurs were like you, they had fully formed lungs, livers, bloodstreams in a fully equal to your circulatory system. Dinosaurs were fully developed in the digestive and the skeletal systems as we can see. The big difference is that they were not intelligent as humans are, they were not conscious in the same way humans are either. Yet in all this we look at their past and know that such events, the advent of human intelligence and consciousness would follow. The intellect and consciousness was not required for the skeletal, digestive, cardio-vascular systems and survival as they had both well under control. Evolution, was itself intelligent, the DNA itself was intelligent. The DNA is a library of nearly limitless possibilities built into its programming. The religious will at this point say, it was engineered or created. They see the god of their respective religions in this. The complex programing that is the DNA is clearly designed by intellect but that does not necessarily implicate a singular grand super god. There are intellectual intermediaries in the Field, amorphous intelligences, entities, busy entities who discovered RNA forming itself lodging in lipid cells. These beings learned to manipulate and program the DNA code. They assembled it for all possible outcomes of things seen, holographic things, in the grand Field of grids.

The potential for consciousness, though it had not manifested in the material realm during the age of dinosaurs, was clearly predetermined, pre-programmed, or pre-present in the DNA. The DNA was programmed and tuned to a variety of multiple environments. The environments are programmed to draw out the features related to the environments thus yielding a large possibility of different creatures. The DNA patterning potential was inspired by the holographic field of creatures seen and unseen.

When I observe all the early animals in my book of fossils it is clear that the world of the flora and the fauna were progressively being developed, tested to see if more and more complex forms could be brought out of the DNA programing and patterning code. It is clearly an experiment which we see results of in the earth. The earth is unmistakably a giant laboratory full of experimentation. Body systems were developed first, stabilized then intelligent consciousness was pulled through in ever gaining increments to manifest its greatest manifestation, mankind. Not only was life forming on earth, it was being developed and driven to improvement. There is no reason that random evolution would move to improve its designs, that would demonstrate intention, intelligence and PRE-Design. There is no missing this observable conclusion. This is not something discussed in anthropology. The reason is that the implications, religious people reflexively leap to conclusions that a god engineered all the life on earth, cutting short empirical observation. Something else is involved. The DNA itself is the CODED, a complex computer signaling program, the library of all digitally coded possibilities. The DNA is clearly a material reflection and patterned on EM radiation, coherent and incoherent. We may not yet have seen all that will be manifested from the DNA.

Pre-existent intelligence and consciousness were in the lines of force before the Big Bang. The bang tangled and tied and tethered lines of force together, this slowed the lines of force in such volume that it created the material realm. Intelligence/consciousness did exist eternally, the material realm has not. Remember to think back to the dominos which represent the variations in size in of our recognizable lines of force. These groupings of lines also a paired and in proximity to smaller and smaller lines of force dimensions which tumble and trigger endless big bangs in static, at rest version of the fields of force. Consciousness in the evolutionary record seems to be pushing through the material of different densities. This is all real difficult to make sense of but then again some of this is just seeing and accepting. One realm causes a big bang in higher, bigger, slower realms above them. Nature is holographic.

In order for evolution to be driving toward advancement it reveals the pre-presence of conscious intelligent, showing a patterning or pre-coding. The fact that evolution ended up manifesting consciousness similar to the programming found in the DNA into our own computer systems indicates that our consciousness is not native to nor dependent on matter and material housing. Consciousness is writing its name in the matter, pushing into the material world, consciousness in all that it was ever going to manifest was written in the E.M. spectrum PRIOR. Matter is slowing consciousness down to a crawl; it's what we observe as "evolution". Consciousness is not native to the material realm, it is slowed down here. Consciousness is "Electrical" but the spectrum is more than just electrical. Matter is manifesting densities and environments composed of line entanglements. E.M. Radiation is purer than matter; E.M. Radiation is interplay of magnetic and dielectric lines.

DNA is not really evolving. However DNA is not done manifesting all of its potential creatures and manifestations of consciousness as conscientiousness is a part of evolution, serving evolutions purposes. What is the ultimate "goal" of evolution? The goal is eternal, infinite digital storage of what was patterned in the slow medium (a.k.a. the material/finite universe). We and all living things of all time have pushed through the matrix of matter creating permanent, indelible digital encodings on the record which is the field as if it was a blank digital compact disc or DVD recording. We are in the movie. We are imprinted, our memories are real. It's odd to think of but consciousness is separate and independent from the material entanglement that made of atoms in the lines. Consciousness pushes through and is inhibited, slowed down and retarded by matter, visualize the scene in the movie "A Nightmare on Elm Street" where Freddy Kruger is pushing into the material world through the wall, so it is with consciousness and evolution. Matter can be thought of as a coagulation of the tangled lines of force in all the various manifestations causing different densities through which the different frequencies of the EM push through differently.

Metaphor and the End of the Universe

The metaphor of the DVD player corresponds to the known universe, the one which our "brilliant" quantum physics scientist's prophecy the doom and ultimate heat death of. The DVD or the magnetic tape or the metaphor of what is recorded is recorded in the lines, the eternal lines.

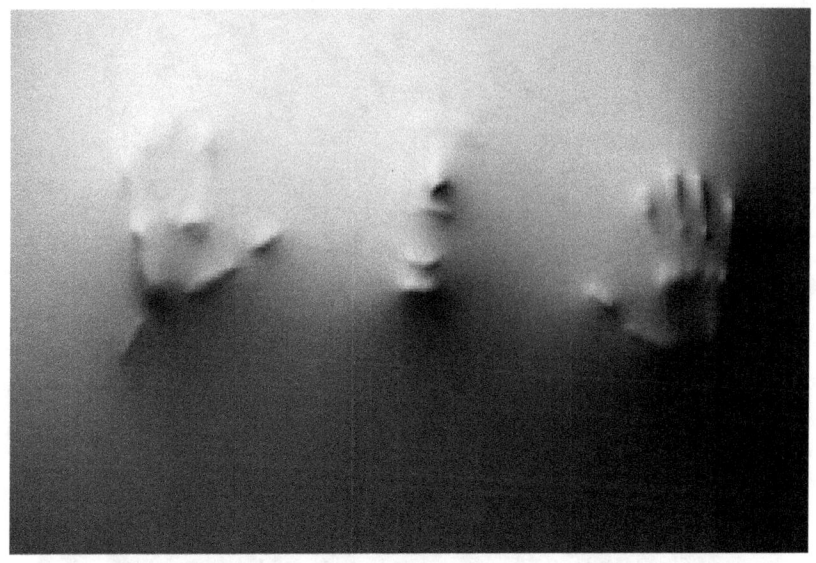

Graphic exemplifies what the EM Spectrum is manifesting through matter, through the material medium of DNA.

Evolution through matter started with a small push into the material, bacteria and amoebas, these were the manifestation of just a finger, then a hand then full conscious forms. Consciousness is the progressive wave pushing linearly through matter, at death we separate the consciousness wave from the material up ending where we were positioned in a motion of time linear movement through the paired lines of force. Consciousness is the domino wave, the big banging wave.

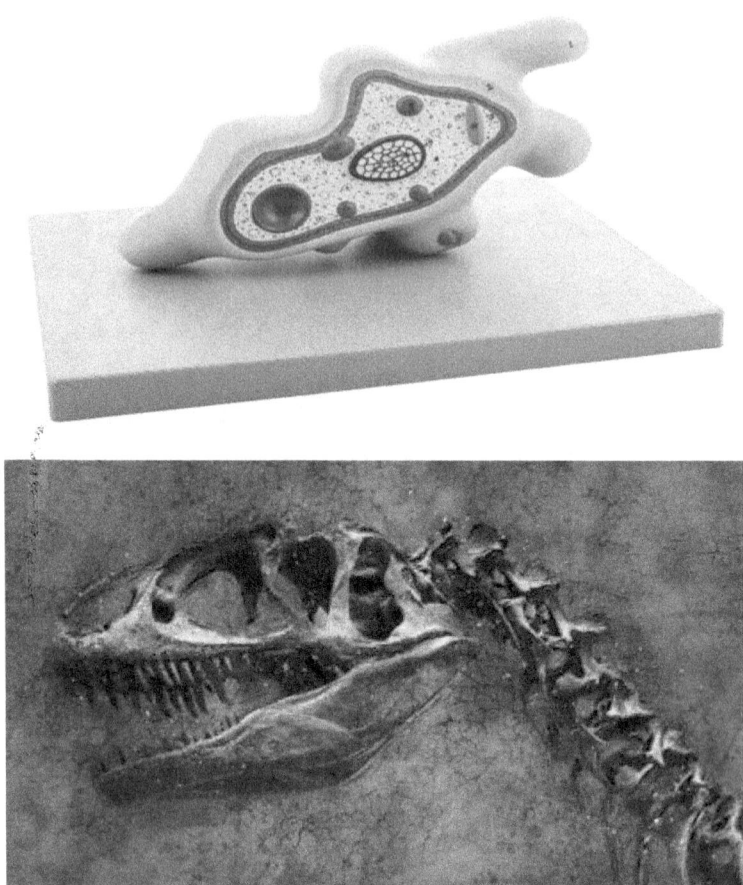

When you see the Amoeba realize the DNA for a dinosaur or a human's construction may have all been there in the Amoeba, in viruses and in "simple" bacteria. Even a carrot has more complex DNA than humans. Humans have approximately 23,000 pairs of chromosomes and a carrot has slightly more than that! This came out of the research into the Human Genome Project.

From amoebas to dinosaurs and on to humans, evolution has been a progression in complex creatures and all that they can do a buildup. Symmetrical limbs have been built up, organ systems, eyes, ways to sense, ways to defend, armor etc. What consciousness has been able to manifest has done so bit by bit in a gradual addition and progression, like a small snowball rolling down a hill it gathers and grows what it is in size and complexity. It is a worthy analogy of evolution. Consciousness has grown its diversity and forced more of its true self up to the surface of the atomic matter. See the illustration below for a visual of evolutions gradual manifestation on the surface of matter which was directionally driving it all along. Evolution is a progressive revealing or manifesting through gathered momentum of what the EM Spectrum IS!

Illustration of directionally driven evolution by the wave of consciousness adding, cumulatively to complexity.

DNA does not equate to the product of creation by a God who is conscious and intelligent. The consciousness and intelligence clearly observed in the DNA is rather a manifestation, a material reflection, of the conscious intelligence which is movement/motion traveling as a wave through the lines of force. In answer to the scientist who denies that the DNA is intelligent and a manifestation of deliberate consciousness just realize that the scientist who would say that, with his great intelligence and consciousness is still unable to decode the DNA with all its trigger manifestations. Such a scientist is unable to replicate any computer program as well designed.

The Field is a repeating set of paired, conjugate lines of force. The lines consist of a magnetic line of force paired to a static line (dielectric) of force. A given magnetic line can cause a static or "dielectric" line of force which in turn can cause a magnetic line of force which in turn causes a static line of force and this goes on at infinitum into infinity with literally no end! What is bizarre is that when the lines interplay, or weave we call this the E.M. Radiation. The field when set in motion, causes light and "electricity" (Forms of E.M. radiation), for example are unmistakably a configuration *mirrored* by (and in the shape of) the DNA code found in all living creatures. Electricity, light and DNA manifest a spiral of a double helix action. This seems to be linked to "brightness", intelligence or even CONSIOUSNESS. E.M. Radiation is literally creating a footprint of its own internal reality with appropriately configured nucleotides! What the nucleotides are manifesting in a strand of DNA is a reflection of the information of an E.M. Radiation stream. How many "world" and creatures, both manifest and un-manifest are patterned in the "light"?

You may never have understood, but any good bio-chemist will know that it is electrical forces acting on nucleotides that first causes them to form and which organizes and dictates the shape of the DNA. Electrical forces cause DNA configuration and order, electrical forces cause the unzipping action of the DNA to replicate itself, the proteins involved are all driven by electrical forces. What has not been stated or properly stated before is how electricity or E.M. Radiation is creating a mirror image of its own unchanging, unevolving shape and coding in a material medium. Like a footprint in sand, so the electrical forces acting in the medium of Atoms and nucleotides, is what the material pattern is an imprint of! Absorb that! The animals which DNA has manifested in the world are actually <u>more alike</u> in their DNA configurations than they different and only manifest different creatures as triggered by environmental consideration. The implications of this are huge, who environments may be digitally encoded in a stream of E.M. Radiation as well as the possible creatures which can and should branch off from these environments so as to gradually manifest them in the slower medium of atoms. Because the medium of atomic matter is so much slower and has a retarding effect on E.M. Radiation only a slow and progressive manifestation was possible of ever more complex creatures was possible, hence visible evolution. This explains why plants and animals show evolution but why DNA does not. DNA came ready and complete, fully formed. This is because it was never "formed", it was not "forming", it was an imprint, a reflection of a very real and static environment or set of environments filled with plants, animals, people and all of recorded history with all of their possibilities already there, static and complete. (Parmenides' theory)

Think about the observable facts about how DNA and the life which it has manifested so quickly on the earth un-mistakenly seems to be self-organizing, designed for a direction and goal with a deliberate end of progression built into its manifesting designs. It is at this point where the less scientific start to default their explanations of why life encodes sophisticated and directionally goal oriented manifestations according to environmental triggers to a "god", or higher power, or space aliens etc. All these conclusions fail to see that the very real and tangible field of forces, has shape, has direction, has motion triggered by imbalance in the form of the golden ratio. In other words, because a portion of the limitless field is larger than other portions of the field by a ratio of 1.618034 to one, this imbalance sets the field in motion like a balance scale with one apple set on the left and then a second apple and a half are added to the right side of the scale which quickly causes the scale to go into motion as it tips right!

Consciousness is shape, shape is motion, motion is patterned, and consciousness, intelligence and directionality are native to the shape of the field. In other words, some may wonder if consciousness (god) manifested (created) the material universe or did consciousness happen as a byproduct of the material manifestation. This is where a study of Aristotle would do a philosopher, a scientist and theologian a great enlightening service. You see, the theory of evolution as put forth by Sir Charles Darwin was a subject of study way back in ancient Greece long before and the particulars of this debate were ironed out with logic by Aristotle back then. Aristotle, Empedocles and Anaximander among others of the ancient Greek philosophers observed that more complex life forms descend from simpler and earlier life forms. This was not in debate or question among these. What was in doubt was weather nature organized the evolutionary process simply by chance as Empedocles surmised. Aristotle demonstrated the fallacy of this idea as nature just happens to provide what is best for advancing species. Aristotle points out that nature is not random at all in what is manifesting but rather it manifests according to densities of the states of matter, which in modern times we designate as *solids* (earth), *liquids* (water), *gases* (air), *plasmas/electricity* (fire) and the *field* (ether). Aristotle points out that these self-separating densities manifest the patterns into living DNA is exposed thus driving the product to manifest a mirror or the densities of the field entanglements.

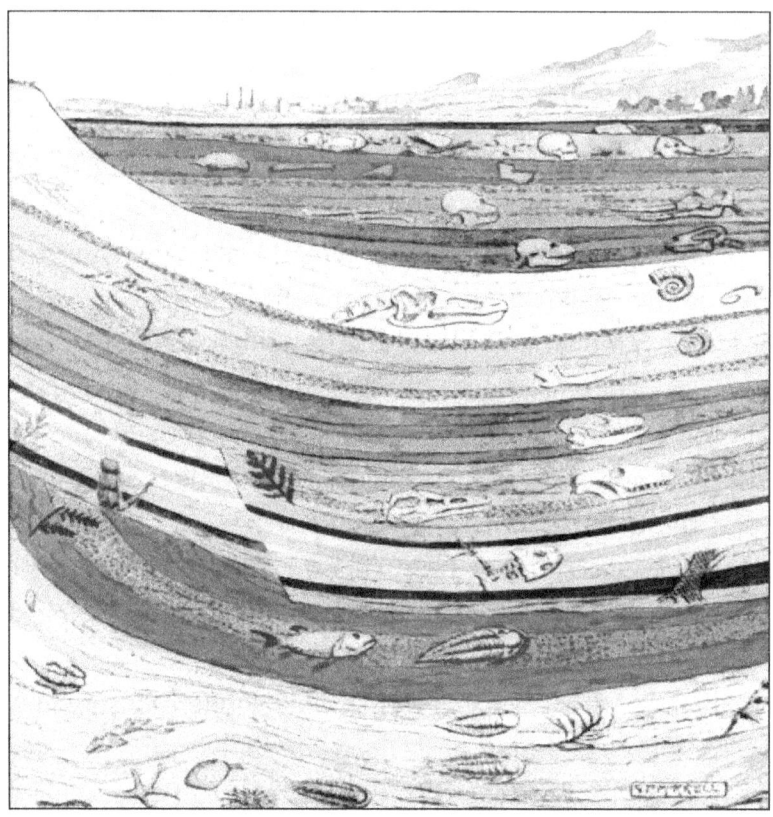

The observations of Charles Darwin, Anaximander, Empedocles were accurate, however Aristotle makes the clarifying observation about the evolution which occurred on the earth. It has a self randomizing, mutation component but that component did not innate evolution nor does it drive it to better adapt species to survive and thrive in different environments. Aristotle figured out that the evolution which occurred on earth is SUPPLIMENTED EVOLUTION. More clever engineering and programming than what we can deliberately accomplish is at work in DNA manifestations on earth. Not possible even with quadrillions of years to do so. Those who were conscious in the field, in the Eather used the naturally occurring nucleotides and RNA strands as programming for biological vehicles used to interact in the matter realm for themselves and for all the living forms we see evidence of.

According to Aristotle's thread of reasoning, human consciousness is an end product driven by the shape of the fields own densities manifestations in matter but not at all limited to matter. DNA stranding is a reflection of the coaxial nature of light and electricity, and what's patterned in the DNA. DNA is not a "creation" engineered, it is a manifestation of inevitability as is human consciousness, patterned on what simply is, what is extant in the shapes and patterns in the eternal infinite Electro-magnetic field. DNA, and evolution are definitely directionally driven and by extension so is intelligence and consciousness. These, are not "creations", these are reflections of the field.

Consciousness in man is a product manifesting evolution; it is a progress advancement in the species forced to the surface from within the already existent parts of the embedded code! Consciousness is brought forth in order for a living animal to better hunt and remember where it put and stored food and where to perceive that other creatures cannot find your hidden food. Yet in saying this we must realize that such a direction is an improvement in the species which is goal oriented to adapt the species to improve to better survive but why is that a goal for a creature that dies after a short life span? Think about it.

Since we have deduced that consciousness is from the E.M. Spectrum and that it had no beginning ever, we must further conclude that consciousness is not a thing with a beginning or any kind of end. Consciousness is not an individual thing as thought of as a thing found in individuals. The self-replicating DNA continues on and on in millions of life forms and is a push into the material world but is one continuous traveling reality through matter over a linear time path.

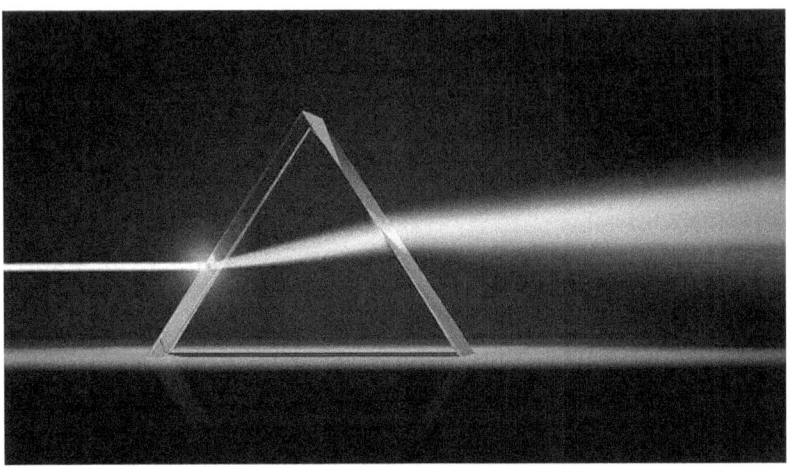

Light prism graphic also illustrates a metaphor of DNA products as an output of Light through matter.

Another analogy/metaphorical way of understanding that DNA is a manifestation of the E.M. Spectrum acting on the atomic/material world can be seen in the graphic of a prism below. In the graphic let the white light entering on the left of the prism represent the E.M. Spectrum in wave motion arriving at the prism. Let the prism represent that which is slowing down the moving spectrum which is full of information. The color light which comes out of the prism represents DNA with its produced diversity range manifesting in the material world, enabling "life". Matter is a matrix which is slowing and parsing the consciousness of the E.M. Spectrum.

Following this thread of logic we can consider as fact then that beings of light both flora and fauna must exist as real imbedded on the E.M. Spectrum in its patterning. The beings of light (E.M. Radiation) are less impeded by the limitations imposed by matter which acts as a mechanism of slowing down light and other E.M. Spectrum radiation. One metaphoric way of understanding the effects of matter on E.M. Spectrum consciousness would be like a runner, ready to really move being made to run in mud which is up to his knees, or like a runner asked to run in bubble gum up to his knees and all around him or her. For really intelligent people trying to motivate and educate less kinetic people this is what progress and innovation feel like to them. Matter can be considered a retardation factor on the otherwise fast free consciousness. As odd as this analysis may seem it begs a quote from Hamlet by William Shakespeare where he correctly wrote "There are more things in heaven and earth, Horatio, than are dreamt of in your philosophy."

Return to Plato's Theory of Forms Pythagoras and Parmenides

The above observations about the products of light and what it is reflecting into the matter/material world brings us to a long discarded theory of Plato student of Socrates of Athens. Plato, in The Republic, stated that the world in which we live is a poor shadowy version of a world which underlays the one we know. There is a realm of reality more perfect than the one we live in. There exists the ideal realm, the digital, the light, the perfect ideal to the forms we know in our common material realm. Plato informs us that that which we consider "real" is poor series of reflections of more ideal forms outside of our perception range. These observations of Plato were arrived at over two thousand years ago using arithmetic and logic, it turns out that Plato was absolutely right and now stands fully vindicated.

Plato in his theory of forms also extended the theory to explain learning, what it is and its connection to the ideal realm. Plato taught that when we learn we are simply remembering what we already knew from the ideal realm. This is a theory which was dismissed out of hand by Aristotle but now, in the new light of modern empirical scientific observations it seems that Plato wins the debate once and for all. The fact that Plato was correct does in no way undermine Aristotle's explanations of prior analytics and how these, in syllogistic form, contribute heavily to how we know what we know.

To be clear, Plato summed it up that we are not in the ideal realm from which was the pattern from which the material realm is an approximation, estimation of. This is not the realm where you will see "Justice" or any other prevail. We are in the realm that retards the ideal and are ultimately overwhelmed by the retarding effect of tangled massed up lines of force. Electromagnetic radiation impacts and runs aground into the small, finite, terminal area of the field which we now identify as material reality. Life and death are a misunderstood thing in the material realm. Life initiates in the DNA form when it is collided with electrically compatible, ideal and reflective atoms and molecules build around the EMR which has in its composition mind boggling volumes of reality, information, ideals and patterning. The DNA is reflective of this and energize it as it passes through the matter. As the electromagnetic radiation passes through the matter it also exits and the DNA shortens, degrades, "dies". The EMR goes on, continues to exist without end.

The features of the life come and go with the and are only temporarily featured in the matter of this world through the DNA.

This is not the realm of the ideals, this is not the realm of the eternal hope, this realm is quantitative or quantum. It has a beginning and end, a life and death cycle and in the end heat death will conquer every last bit of the material universe. Don't expect to win in this realm. You can, you could fall into a pocket or good fortune but this will be exception instead of the rule as a whole. The Book of the Earth tells the same story that Physics and electrical studies show us. Hope and gods etc. do not come down to the earth to rescue us. The only thing coming is a comet to wipe life off the planet as it has at least twice before. Smaller comets may have cause the other extinctions and ice ages etc.

This is the analysis and observational sum of consciousness, analyzing what it even is. Consciousness is NOT something which can be properly understood by means of the disciplines of biology, biochemistry nor psychology as the source of consciousness is ties back to the Big Bang and Electromagnetic radiation. With this understanding about consciousness and its source comes immense implications which answer age old philosophical questions about predestination, metaphysical, religious and other behavior material questions. I hope this opened a few eyes to the nature of nature and reality! Consider still further the implications of the dream state under this understanding of consciousness. People think dreams are "not real" but in line with where consciousness originates from and where it goes at the end of our bodily life, we may well realize that the dream state is also a departure from bodily confinement. A dream state may very well be a close parallel to what consciousness we may experience upon the breakdown of our bodies.

CHAPTER 8

The Conclusions, the Field the Sum Total

In conclusion, beginning with Thales we have traveled down a line of scientific observations and reasoning over a course of more than two thousand six-hundred years. The majority and the best of, those thinkers and observers over the years almost all came to the same conclusions that reality manifests as a Field. One grand field exists, a field within fields and downward descend field at infinitum. Such a field exists; the field implies endless depth at infinitum in layers descending in a ratio of 1 to *phi*, 1.618034. One set of lines is 1.618034 in ratio and size to the next downward set of lines which are one compared. This line which is one in ratio is in turn, 1.618034 in ratio to the next downward down descending line and on and on forever! This is the field, in size ratio there are same size pairs of conjugate lines, one can be visualized as a magnetic line pitching left, paired to another same sized dielectric line pitching right in descending ratios forever. The truth of this reality was only speculated and assumed by rudimentary observations in ancient times.

Later, after a large period of western society loss of information a renaissance happened and much of the ancient wisdom was recovered and expanded upon by observation. This led up to and began to overlap the most piercing age of scientific facts, observations and discoveries reaching up and into the twentieth century. The field of speculation and rudimentary observation was found, used, manipulated for the huge advancement of the electrical sciences. The sciences were producing electrical forces and grids and wireless technologies which seemed to contradict the operational model of particle physics. The rise of relativity snapped back and pushed the field science into the corner and replaced it with the atomism of old being repackaged, reborn and re-sold to the mindless public. These were not mentally equipped to see the bait and switch inferior imitation of the field in the form of "speed of light", and "time-space" continuum, fabric of space propaganda was a trick. Never the less there are those of us who pushed through the smoke screen to rediscover the field and to experiment with the lines of force and study the old thinkers and their miracles of math which explain the field's behavior and mandatory collapse and sudden expansion (Bang). True science has fallen out of favor in the main stream but is now Avant guard once again.

We examined just how particles of atoms are tangled lines in various combinations of disposition and size. Electrons, Protons and Neutrons are all different ratios of magnetic and dielectric entanglements and E.M. radiation is the product of braided magnetic and dielectric lines of force. The two manifestations of the bang encounter one another, the faster manifestation clothed itself so to speak in the slower causing a two part reality split. One that just is fully formed and unchanging and one breaking down, forming, dying, decaying, recombinant over and over as long as the material manifestation lasts. This is the Parmenides and Heraclitus paradox which was resolved in Plato's theory of forms.

Einstein's spooky action at a distance was resolved in the connecting of the evidence of the field back to the proper understanding of the field. Quantifiable particles in a vacuum can never explain spooky I.A.A.D. but not for a lack of patchwork and trying by the stubborn particle physicists. We also read of Isaac Newton and what he really had to say about Gravity vs. what his sycophants represent his understanding to be. Isaac knew that the bodies in space were not unattached in a void, as we perceive them to be. Still he did not know about electrical forces and how they act on bodies due to their tethering through magnetic and dielectric lines of force in a field. Electron flow could never explain I.A.A.D. How the facts became buried under mountains of disinformation and nonsense is unbelievable yet truly what has happened to our science! It's time to spread this world to those smart enough to see its value.

What has been explored in this work is an all-encompassing natural philosophy which answers the puzzles of Physics, Metaphysics, Psychology, Chemistry, Biology, Philosophy and even Religion. In the spirit of Thales I did not stop at superstitious religious revelations. We must understand what religion is. Religion began as a very primitive version of science. Religion began as a system of answering questions which an evolving race was becoming aware of. Religion fit a niche of supplying answers to questions of who am I? Why am I able to ask such a question? Does it mean anything that I can see and think? If so why do I die as if nothing I did mattered? Is there ultimate justice? Where does that justice come from? Who or what regulates justice? Is there equality somewhere beyond our earthly life as life is clearly unfair? This leads to the question of, is there authority over justice, a gods or one super, overarching God? Even our great philosophers grappled with these questions and fell short in some regards.

Let's start with the moral law, In the Jewish and Christian religions the ten-commandments are the centerpiece of their moral codes of law. But moral law is not strictly a religious concern as any successful society has to have and enforce a moral code of laws. What is fascinating is that close versions of ten commandments pervade the legal codes of all human societies as such codes are realizations of what society needs to function well together as it serves as a common social contract. The ten-commandments are a more dedicated religious version of the same generally agreed upon laws. A set of laws should be retained by us even though we find that none of the established and traditional religions of the world got us to our understanding of the great FIELD. The long march through time to arrive at these truths of nature and reality were accomplished AROUND and in spite of religious impedance. As such religions and the false narratives which they have been clouding our understanding and research for millennia should largely be dismissed. With that said it is foolish to completely dispense with every bit of religion as many have been the keepers of the moral codes which have formed a justice system for their people. That is their main value. Recognizing the God versions very from Islam, to Christianity to Judaism to Native American religions we need to modify our moral code of ten commandments to something much more like the following...

Ten Moral Absolute Laws to be agreed with

1. The compounding manifestation resulting in consciousness brought you out of ignorance and superstition; you shall harmonize with the facts of reality, as best can be understood by you, with all of your emotion, with all your

intellect and with all your total being. You shall have no other higher ethics than these.

2. You shall not deliberately formulate and propagate nor glorify any false models of reality (pet theories/ignorant opinions/Idols).

3. You shall not falsely represent the facts of neither reality nor slander or commit liable about those who research and teach about it.

4. It is the moral obligation of those in charge of others to allow and offer frequent and periodic reposes for working men, women, children, animals and even mechanical devices, at least for maintenance. Periodic breaks will be scheduled daily, weekly, and monthly and annually out of a reasonable moral sympathy and empathy for your fellow beings.

5. Show a reasonable minimum of love, honor and respect to your father and your mother (whether they merit such treatment or not) and to the parents of other people also.

6. You will not commit homicide, which does not include accidental manslaughter, or reasonable service in moral war which is conducted for the survival defense of one's own nation and immediate family even down to a family of one.

7. You shall not shame another man, or a woman, or their children to secretly lay sexually with his or her spouse without the full consent of both parties of the marriage. No two-party consent is acceptable either when minor children are present in the same home.

8. You shall not steal; take what rightfully belongs to someone else without their voluntary consent.

9. You shall not give false witness, claim that you witnessed what you did not, nor shall you omit or deny, or otherwise edit, what was witnessed in matters of public trust with others.
10. You shall not jealously crave to have the possessions that other people have. Get, make or build your own desired possessions.

With these laws set in order to reflect the natural philosophy and the needed order of human society we come to other questions related to religion. The question of God will come up when someone reads this work. The God of all the major religions is NOT neither who nor what created this reality as it was expounded in this work. This does not mean that there is not such a being. If there is a God this God has to fit in properly with what has been discovered in the research of the Field. The field, an eternal conjugate of magnetic and dielectric lines of endless dimensional descent has a wave passing through it and traveling through it. The wave when impacts the region causes collapse and expansion due to conjugate tumbling of different ratio sized lines of force. There was no spontaneous "creation", not even of the very old DNA strand as it formed when it could around and patterned after E.M. radiation of the right proportions in the soup of adequate nucleotides. If one wants to say that God is the traveling wave in the FIELD well there is room for that claim. On the reverse of this thought it wouldn't do you or anyone much good to pray to Electro-Magnetic Radiation, nor is it safe if you were to get into the presence of ultraviolet-rays, x-rays, gamma-rays, cosmic-rays etc. It wouldn't do anything to pray to the phenomenon. EVERYTHING depends on definition, how big is your definition of God? The Bible of the Judeo-Christian texts claims many broad definitions of God, one of them being that "God is Light". By such a broad overreaching definition, if we make and accept that equation and definition, then by that definition God has to exist since Light (E.M. Radiation) clearly exists. This is an odd definition to assign to the God written of in all the religions of man however as the religions which were built around a God idea clearly contradict one another and lose all credibility as the authors of the so called representative works have all "born false witness" regarding the origins of their authority and fought all of these discovered facts uncovered by science of how we all came to be here! Also bear in mind that "light" in all of its forms of Electromagnetic Radiation is a minority feature of the boundless, conjugate field. This definition does not match the Omni-present designation of all the religions of humankind.

There is the question of evil, devils or a grand devil. All I can say to this evil as a conjugate is a natural law and imperative where "goodness" is extreme or concentrated in one's person or location. Anything pulled out of, an at rest state, will become energized through tension which will cause a pull of energy to a relatively empty location, like electricity. With that said, the concept and manifestation of evil is a distortion, an aberration, an abnormality as is being too good as nature implies a balance of forces. Evil is not necessarily the opposite of good as good is average, calm, normal and balanced but evil is the opposite of the distortion of hyper, unrealistic righteousness. Does the phenomenon of "evil" manifest real coagulations at times, in people and at certain places? Coagulations of hyper-righteousness and "evil" are oddities but seen in the world at times.

There next follows the question of the afterlife, heaven and hell, Elysium, Valhalla Hades etc. The only honest answer that I could venture with credibility has to be given with reference of what is known about the Field and E.M. Radiation. The light, the radiations (electrical forces) form the DNA around them. The universe, the worlds within the E.M. Streams and all the life forms and environments encoded in the beams are a mystery, completely unknown to us. What is digitally present in the light is similar to what we see it cause to exist in the material world but we need to realize that what we see now and in the fossil record likely only represents a small fraction of what is possible to manifest from the code. The creatures which have manifested, including humans, were forced up or teased out of the digital contents of the generic DNA by multitudes of variations of environments. Think about what this implies, this implies that the animals are not just coded by the DNA but all possible natural environments which correspond to those animal variations may also be fully present in the streams of light around which nucleotides form. This may factor into life and death destinations as the DNA is a reflection of Intelligent Consciousness beyond the ability of the brightest programmers of today have to produce. The light was before the DNA and pushed through matter to gradually manifest, little by little more complexity and intelligence and consciousness in its natural product. As we age and DNA shortens it represents the light lattice pulling apart and away. This is the slow death outcome. Consciousness is a thing very different from what biology and psychology have understood it to even be. This means that what we call "life" is also completely misunderstood unless we understand the nature of E.M. Radiation in all of its manifestations! DNA, unzipping from life equals what we have called "death", and the opposite, the forming of "life" is defined as the zipping up of nucleotides to a light, Electromagnetic Radiation scaffold or frame! Life and death that is what it is. Consider what was just stated in the light of an extensive life after death study where thousands of declared dead patients (Two thousand specifically) in emergency rooms etc. tell the same story over and over of consciousness remaining and other environments manifest for the consciousness. These cases have been carefully studied by real medical scientists and doctors over decades of time. The most extensive life after death study was conducted in the U.K. at the University of Southampton over four years. These medical researchers studied cases in the U.K., The United States and in Austria at fifteen different hospitals. In this study forty percent of the people that survived reported having some form of awareness after being declared clinically dead. The revived, almost all tell of seeing a bright "light", not of this dimension, light is a universal sight regardless of what religion or lack of religion the person claims.

Factoring for Time Displacement

Another set of facts concerning the Field which agree with life after death scenarios is the descriptions of people somehow seeing more and understanding all things much better than they did in life, in their body. As we explored concerning the Field, the field is static and fast acting and fully formed, containing all of our history before it happens yet matter, the atomic manifestations are slower, acting as a retarding (slowing) factor in the formation of DNA. The Big Bang was 13.8 billion years ago but DNA was only finally seen and manifested on earth a mere 2 billion years ago. This implies that

DNA was not able to form for 11.8 billion years, it was significantly slowed, impeded, retarded for all that time. A more conservative approach would be to take the age of the earth, 4.543 Billion years and then deduct the two billion years of life on it to see that it took 2.543 Billion years for DNA to form up on earth. Our material and atomic life spans and bodies represent the slowing version of intellect and consciousness as it is caught or retarded in a relatively slow moving medium. The further observable information verifying this same point is that even though DNA formed 2 Billion years ago it has only been able to evolve up into manifestation creatures nearly as intelligent and conscious enough to replicate DNA programing only as recently as maybe 10,000 years. This is a slow, slow, slow realization and manifestation of intellect on the earth. "Life after death" people seem to be describing the same thing in conscious humans as we have seen in the DNA track of evolution. Conscious thought is a product of a higher faster bandwidth than that of the material, tangled lines of force realm. This is why consciousness in simile is like a bullet slowed down when shot into water, or into ballistics gel, this is why so many "life after death" experiences describe a time displacement of being dead for minutes but while dead experiencing scenes which seemed to be for days or longer.

E.M. Radiation, the true nature of matter and the resting field is a set of mysteries and science worthy of us exploring much deeper than ever before, we need to rejoin the research where it was left to die. A recrudescence to true scientific investigation into the nature of the natural order is long overdue!

To summarize, we are not describing souls, gods, angels and devils, we are not examining anyone's versions of heaven and hell. "Life" as we observe it, know it, and experience it is a formation of nucleotides in a specific order and lattice structure, PROMPTED TO DO SO, UNQUESTIONABLY by "electrical forces". There is literally a zipper, zipping up action occurring due to electrical forces, but what has been poorly understood is just what size or type of "electrical forces" are and what they look like. The DNA manifestation is built on electrical forces which are on a scale of either ultraviolet or smaller. The electrical forces have a separate physical manifestation apart from the molecules, from the nucleotides but classical chemistry and physics will not reveal this nor find this reality because the electrical sciences were shut down after Tesla and Steinmetz. The intelligence and consciousness is in the light, the radiation, it is intelligence and consciousness, when it departs it departs from the matter matrix which it is zipped up with. At "death" the Electromagnetic radiation remains intact and departs, the nucleotides degrade and break apart (decompose) their structures and such molecules are set free to be joined to something else. The radiation goes on with its path which it was on before one was ever born. Consider if you will, how almost instinctively how humans describe very intelligent and very conscious and conscientious people as being "bright" and "brilliant". In fact it is our most brilliant minds that invented the designations of bright and brilliant. How many of us describe when others finally learn something as having seen the light or the light has finally gone on? This is not an accident; this is the light communicating with our retarded brains. Yes all of us are retarded, this is not a slur, this is literal, and retardation is an obstruction to full flow of thought, a barrier or set of barriers or barricade. Our nucleotide, material, matter brains, composed of cells and atoms formed of entangled lines forming the mass of atoms in fact does slow down light exactly like a prism slows down pure white light which passes through it. It is a perfect analogy.

In full circle we return to our great original Greek philosophers and we find that so many were right, especially when taken together. Parmenides was right in his conclusions that indeed, no one is born, no one dies, no one lives out a life time in the grand total of the static state of what is recorded or imprinted (or just is) in the complex structure of light (E.M. Radiation). Were we ever here? Yes, but our senses did not reveal this model of reality, our senses were part of the material, matter, atomic reality and therefore a deception trap which tricks us from gathering the fuller picture of reality into which the material fits. Once light was zipped to the molecules to the nucleotide formations then indeed a time linear pattern begins to cycle begins to be set in motion which will indeed form replicating cells, dividing cells, forming up into a fetus, then a birth happens, a life time, a departing of the light lattice etc. But the light, the E.M. Radiation there is no linear time and movement through time.

Parmenides was on target concerning the reality which originates in the light but this does not mean that the product of the tension which both Heraclites and Empedocles observed was wrong, the material (atomic and sub-atomic) was in fact generated from two opposed forces which we call magnetism and di-electricity. Heraclites called the creative engine "tension", "Opposition" and "war". War was the oppositional creative force. This is much like the big bang first described by Empedocles so long ago calling the opposing forces "love and strife". Call the lines of force what you like, these remain undisturbed and at rest in the larger "part" of the Sisyphean field of POTENTIAL. Imbalance in the lines, or in the line chunks, keeps the whole field in motion in Perpetuity! The imbalance gives shape, oppositional shape; shape is motion/movement and time. The main perpetual motion machine is the field itself and we and all that we can conceive of as reality to our sciences, from the "macro" to the "sub-atomic" all fit in only one line, or line grouping of force in a field which is boundless in size and proportion. The day has come for a recrudescence to true natural philosophy's investigation of the true nature of reality. It is long overdue and needs to rise in ascendancy!

To summarize what I have uncovered here in adumbration, the field which was previously known as the Aether, is what now could be understood to be the "Matrix". The Matrix in the movie the Matrix was a binary computer generated world. Our field is also a Sisyphean/recursive boundless field of juxtaposed and binary lines stabilized and at rest at low tension and integrated and tensed to produce matter and energy in all the states which we can identify. The field contains all that has been and will ever be, it is all in all, one field, binary in nature forming the matrix of all which we call reality. Computer programming mirrors reality!

God, Aliens, Demons, Angels and Other Misunderstandings of Reality

Like Descartes we must throw out ALL of our beliefs and definitions and then examine each one individually in the light of what we have learned by scientific investigation. To begin with, the

191

idea of one super God, which is the head of all else in the "spiritual" realm is glossily provincial in the light of magnitude of reality examined in this work. God as defined by sacred texts is a small, a diminutive figure, next to the definition of the field. The field is a Cartesian like x,y,z grid which in each square of the grid has concentric scaled down versions of the same grid recursively descending in groups and chunks downward at infinitum, recursively in an asymptotic and Sisyphean manner without end. The field is NOT conscious itself, yet it causes and contains consciousness due to its imbalance of size relative to its different sized line groupings. A small capricious god who out of capriciousness created the spirit realm and the material universe is such a limited being. Such a god would only become god after his/her/its acts of first creation. By definition god was god of nothing before he/she/it created something.

This is not the case for the field as laid out in this work. The field is irregular and uneven in size which necessitates that it collapses by concentric sections once contacted by a smaller high energy section of the field tumbling into the next largest grouping of lines. This causes magnetic and dialectric collapse which pulls in that section of field in onto itself which when compression is at a maximum naturally causes a massive push out which is tantamount to an explosion. This explosion is only half of the action being observed by modern physicists. They are confused about how the Big Banged material can be accelerating. They fail to see the field, the fabric, the grid. If they saw it and understood it, then they would see that the Big Bang is more than just an explosion out from one point (singularity). What was pulled in and compressed was pulled in and down from fixed points in an "invisible" grid far from the singularity point, so that when the compressed lines blew out they also pulled back their stretched and stressed lines of force back out to their previously at rest positions out in the farthest reaches of "space". This is so much bigger than the idea of god. This bang happened at our level, our range of the grid where we perceive micro and macro ends. Below us is a smaller, higher energy sets of lines clustered as a reality which was banged out just as ours was. This happened in a lower band cluster before this without beginning!

God, in order to be the credited conscious former and shaper of what was banged or tangled into existence would make God very small next to the tumble of consciousness which exists in the field. The field as a whole has a conscious and unconscious component to it. Is it whole is it unified? Does it hear any creature in the field? Unknown. The field shapes reality and consciousness but also is shaped by consciousness. If human primitive impulse desires to call this God I would state that this is very short sighted.

Consciousness occurs in the field, perhaps in total, collectively and as a whole and definitely as individual consciousness within beings such as it is with us. We ARE NOT exclusively material beings. In the lines of force, in the "light", or in the electro-magnetic and magneto dialectric braids are infinite tiny faces of reality entangled with consciousness at a range of variations. The beings in the light streams, the EMR/MDE braids are not DNA based life and are real, ever present and self-aware by degrees. They are, and were, something like living conscious animated holograms. Some of those of our primitive race called these beings "Spirits", "gods", "angels" and "demons" etc. depending on their allegiances, if they choose to align with any. Some of the beings, billions of years ago found that they were conscious, self-aware and became competitive and cooperative by degree. These realized that they could effect and manipulate material reality, matter based reality, reality consisting of tangled lines of force, slow and static compared to their experience. Some fraction of what is aware in the field, in the lines of force, made vessels, genetic, DNA replicable vehicles for themselves and for some of the other creatures from the EMR realm. DNA was used, nucleotides assembled to work in a way to self-replicate when set in motion by making contact with Field extant, EMR presences. The non-corporal beings had and have an intellect beyond our ability to participate in, while in this matter (atomic) realm.

UFOs

Do you believe in UFOs? Most people do now. For decades people who have seen and reported seeing UFOs have been marginalized as nuts. They have been ridiculed and ostracized by the cynical side of mainstream society. Then, suddenly in June of 2021 the United States government via the Pentagon released a report which states that UFOs are very real and they released footage of fighter pilot interaction with them. The U.S. and allied force pilots in World War Two reported seeing them as well. They called them *Foo Fighters* at that time. Now if the craft are real then so are their pilots and their engineers. Who and what are they? This is a secret. The U.S. government and other governments think they know. The story goes is that they have had interactions with these beings since World War Two. Is it true? Who cares? You see the government knows more than the public does about this, but We,

The (Conscious) People *can know more than them* and all their cadre of gaslighting departments. We can know who these beings are because we know the nature of the true science, the Aether/or the Field, the bottomless field of recursive energy grids. These beings fit into that order.

Why Entities Don't Just Reveal Themselves In Spite of Government's Suppression

We can know who and what they are <u>through logic</u> and knowledge of the nature of the Field. The logic involved here is similar to that employed by Plato in the Republic when he wrote The Theory of Forms. We already know more than what we realize about the UFO entities if we just think it through. From leaked government documents and anecdotes by key witnesses, we learn that the entities as whole, as a group, ONLY <u>interact with governments, "leadership", not the general population</u>. Think about this, for years we have been told that our governments know who and what these beings are but won't tell the general public. This in turn makes many people resentful for being excluded from the loop of information. Logic should inform you that something is amiss with this explanation of events. Whether the governments know or don't know who and what these beings are, you have to ask, "Do these beings know who they themselves are?" Logic's answers is "yes" they do. If they themselves know who they are and that they have superior technology, then at any given time, they could make themselves known directly to the general public, <u>without any government's approval</u> and there is nothing any government could do to stop them! Why don't they? Why doesn't this play out this way?

The latter question is the million dollar question. What can our government do to stop the revelation of UFO beings and UFO tech if these beings choose to reveal themselves? The answer of course is, nothing, not a damned thing! So keep applying logic, if the above is the case why doesn't such a revelation happen? We are told, such a revelation has occurred at the upper levels of governments, just like it always has. Logic screams at us, that a conspiracy of silence and control over the population exists in this matter between upper echelons of governments, good ones, bad ones and all in between ones and UFO entities. Logic also screams at us that the purpose of the secrecy in this case is the same reason for the majority of secrecy for everything. Someone, or someone(s), are benefitted by the secret at the expense of those kept apart from the information and the truth. Someone is being used. The Romans had sayings for government arrangements such as these, they said "Qui malo?" and "Qui bono?". "Who benefits?" and "Who loses out?" from a conspiracy of silence? These are the questions that when answered will reveal much.

Some who wish to divert you off the trail will tell you that the secrecy of what these beings are must be maintained because that is a necessary secrecy. It is a matter of national security and public safety. This of course is an insulting lie. The upper government officials are those who have contact and a relationship with the UFO entities. This is not new, this is the same old arrangement! This is how it was in ancient Peru, in the Mayan and Egyptian hieroglyphics as well. The kings and their court had direct contact with "the gods" and were the intermediaries between god and man. They became the entitled intermediaries with all the benefits. The public, under the king, had a roll to fear, serve, obey and follow

196

orders. We, just like they, were kept in ignorance, to be used as labor largely through means of religion and patriotic pride. As it was then so it is now. The same old formula is still used today because it works.

The Beings Exist, They are Part of the Natural Order/The Aether, They are NOT "Aliens", Logic Explains Who and What They Are.

It is ignorant to call beings which are involved in earth's oldest formations of life, "Aliens". They were here first, before humans, before animals before plants, even before bacteria. They caused what we think of as DNA based "life" on earth and their history is rooted in and tied to earth. Earth is their lab, they are not confined to earth only but they are centered here, largely in the oceans. They are not what they seem to be *physically*, what they let us see and what they manifest. The beings in bodies, identified by the clueless as "aliens", existed here first. Earth is the laboratory of experimentation, the petri dish of non-corporeal conscious entities that are both in <u>competition</u> and <u>cooperation</u> with one another.

The Entities Did Not Evolve as Biological DNA based life; They Evolved Us but Had a Template

Yes we and all life here is here is a result of evolution which we discovered in geologic time scale but that evolution was NOT random. We were evolved up, we did not evolve as is by random chance. The entities did NOT evolve, not the way our forms did, not from DNA. Not through DNA. They did not come into this material "reality" the way we did. They evolved by a natural process of electromagnetic tumbling or collapsing which occurs in the Field. The tumbling in the Field is much like that which occurs in an

197

electrically charged industrial, car lifting iron magnet which displaces the dielectric portion of the static field. The magnetism is then enhanced, unbalanced to be the stronger force when the electric Field is displaced by overcharging and unbalancing the dielectric part of the Field. This tumbling is natural and has been occurring all along, this is the cause what we call the Big Bang, a collapsing of a section of the larger electromagnetic Field in on itself. When the field snapped back to balance it both blew out and pulled back or stretched back into place causing the tangling. The Field appears to be partitioned by cluster groupings and relative voids, then smaller lines, then relative void on and on downward forever. The breaching and crossing over the void produces collapse or tumbling of electromagnetic lines. These lines and this action tumbling upward in the field is the very essence of thought, intellect and consciousness, i.e. the entities, essentially eternals. They are the amorphous elder children of the Field, the ethereal ones, we are the younger children of the Field in their hands, also from the Field but with forms on both sides.

In our study of DNA based evolution we have utterly failed to find certain missing links. We always will fail until we unite all the pieces of this grand puzzle of the Field. The entities are the "missing link". It is they who manipulated and bridged spans and gaps in the DNA of earth. It is they who turned on and off the premade DNA switches and proteins which shape all the forms we see on earth. They made and programmed the program (software) we call biological DNA, they apply a little radiation here and there and get the results desired to give form in the material world to beings they saw on the Aether side, the holographic side of the Field. We owe the beings everything and nothing. We are here by their actions but they do not own us and have no right to deprive us from rising to the top, this is a universal constant which they are suppressing and violating by working and communicating with unjust and corrupted governments. Said governments are not headed by the best of us. Governments are allowed and aided in rising to the top over the people.

The entities misrepresented as *aliens*, angels, demons, gods, or even *The* one God, did not evolve like we did. Their corporeal bodily forms which have been seen did not come to the recognized forms by a gradual process over millions of years of random DNA self-organization. These beings engineered DNA to self-replicate vessels for forms seen in the Field, complex biological machines. The entities bodies are not like the DNA forms they engineered with the digital, base two, binary bio-programing code as they do not reproduce sexually. They are not male and female and they don't reproduce or age. These non-corporeal entities know how to join and manipulate their own DNA with our DNA and that of animal forms if they choose to. They, as manifestations in the Grid or Field, made for themselves, material, DNA built bodies, not gender nor procreation dependent, like most of the animal life of earth is now which requires COOPERATION by the sexes and COMPETITION against other species to survive off of killing others. They existed in the Field, intelligent and conscious as bi-product of the field's energy and action. They wanted to interact with the slow, material realm. They wanted to use it as an incubator to mature younger beings of intellect in the Aether, you and I. They found an abundance of programmable building blocks, nucleotides and lipids on a planet we call earth. They built avatar bodies as well as bacteria and other simple life forms from the readily abundant components which could be easily assembled and programmed, to make biological machines which could self-replicate and self-perpetuate. They did not make metal bodies, like robots before they made biologically based bodies, as mining metals requires intense physical labor and extraction from the slower matter based portion of the Field. So all metal and plastic machines came later, as they learned to manipulate matter. Just a portion of their reality moved into the material reality, they did not limit themselves by their corporeal bodies. They are not beholden to, nor as physically tied to, nor invested in, the biological orders of the Earth. They do not fear death, they are life apart from biological life, their bodies are only machines. They have physical bodies but these are more like vehicles, bio-robots to ride in and work in and by which they can interact with the solid material realm more easily and directly. This is our reality too but we don't remember coming from out of the Field. We don't remember our consciousness and intellect. Why?

Our Brain is What Makes Us Ostensibly Smart, or Is It?

The answer to this question is like so many other things in this book, it is both a "yes" and "no". Our brains were evolved up to what they are now, to be conscious and intelligent and they are. We among ourselves recognized that some people are gifted, higher in intellect and higher in ethics and ambition and responsibility compared to the general lower functioning public. This is due to our brains, our brains, when worked, worked very hard, and applied, can render us very intelligent and capable. Our brains are capable of elevating, elevating and being educated to a height comparable to the entities that brought us to this realm.

When people die and are revived, it is common to them to say that on the other side they knew more, everything, there was clarity. Our brains that we use here are the impedance, they retard our fuller intellect but we think the opposite because we see people who work those brains seem so much brighter than other people.

Brains Built for Work and Obedience through Religion

Early human writings cast the bodiless entities as gods and we their worshippers, servants. Their ideological interactions and intentions for us, when forming us, caused an ethical dispute among them regarding what is our role and ultimate fate should be in the material realm. Their dispute is one which says that humans have a potential as high as theirs but most humans are content and lazy as beasts of burden who don't want to reach for their potential. Some work to enlighten and elevate us to what we were before in the Field but some of us are older and more evolved and mature while others are lazy and immature.

The great ethics debate of the entities came down to us as the battle between good and evil in most of our holy books. Their battle is a proxy war through us which will go the way the prophetic books say it will. Our destiny will culminate in a battle of ethics. Entities programmed DNA on earth, in the cyanobacteria on the anaerobic earth to encode most of earth's g-gnomes. These beings manipulated and teased out an endless variety of life forms from their handy work, DNA programming, which manifested in the material realm. Life will not "evolve" or self-assemble without their orchestrating and engineering action. We humans divide along the lines of their conflict. Their conflict originates and is based around human creation.

One faction wanted a worker race, intelligent enough to do some complex work but devoid of independence while another faction wished to elevate us above mere beasts of burden to be used by them as we were more in the Field. They are not fully callous, they also want to take care of us as protectors and providers. Another faction was Promethean and wanted to give humanity literal and metaphorical "fire", a.k.a. knowledge, science, logic, technology, philosophy and philology to elevate us to a limitless god like race. Both factions fought proxy wars with each other, through us. This is the war we see now, the tug of war between enlightenment and animal stupidity and selfishness which has no higher purpose than to work like an animal strapped to a plow and be rewarded with the urges to feast, fornicate and fart and nothing more. To describe the human "race" as a, race, is truly a telling double entendre as the humanity is in a race. Humanity has a short window of time to do what it takes to survive as an intelligent inter planetary and interdimensional species or to waste the finite time and opportunity in favor of being just beasts to die in an extinction event like so many other species before us have.

The ethics battle is a logical one though. The formless entities are intelligent as they played a parental, care taker role over beings from the Field. The problem is that the entities threw two very different levels of people/beings into a pool together. Some were older and some younger but of the same type. Some were as high or higher than the entities in potential but none had control and power over the material real. They did and as such they took it upon themselves to encapsulate beings such as ourselves from the Field to live and grow and experience the material realm. Some saw that most humans don't try, don't grow, don't seek enlightenment and philosophy while a significant portion do. As a compromise to a dilemma the entities separated the lower more immature humans from the rulers as intermediaries between themselves and the people. The problem with this is some of the unworthy rose to positions of power and the system became corrupt and did not allow proper self-sorting. Now many intellectuals are abused and held down by economic traps etc.

The entities have agendas, some have a passive, proxy approach when it comes to influencing our ultimate outcome. These beings are a product of braided lines of force, of the FIELD that tumbled and tumbled, twisted and stacked and layered upon itself. The product for them was the layering upon layering, stacking and compacting, forming consciousness and intellect FASTER than that of our own. Their native form is formless, they are the essence of consciousness and intelligence. Their native home is in the Field.

The entities serve their own agendas. It is rumored that they deal with governments as opposed to individuals, gifted or not. This may be due to how they once set up the rulers/kings etc. as intermediaries between the people and themselves. They won't help us much, as we, all the flora and fauna of Earth, are their garden, their experiment. They are not actually superior to the best of us. They seem to be but this is only while we are corporally confined, what our DNA constructed bodies/forms are, is housing for beings from the light/EMR reality. The entities have the advantage of eons of consciousness. They are for all intents and purposes, eternal. They had an origin of consciousness and a then became more and more but so did some of us. We are beings from the Field as well. We came to the material plane, born to be housed in human bodies, due to their manipulations of nucleotides our mirror forms exist here. Life is not what you have come to believe it is and death is not what you think it is either.

We don't have the advantages to know anything from before our lifetimes from memory. Our intellect being plugged into these avatar bodies is impeded and only by application of concentration, effort and concern can we regain and learn back again a fraction of what our entity friends know. While we are here we are confined by the physical limitations of the tangled, retarding/impeding, slowing flow of EMR manifestations of presence. Those beings have most of the human race at a great disadvantage but some part of them have given us fire in opposition to the competing faction. That faction punishes us who would ascend through the weight of oppression to keep us down and animal like but sort of like pets. This is the opposition as told of in the story of Prometheus how Zeus, the suppressor of mankind, sent an eagle to peck out Prometheus' liver for giving us fire.

The fact that some of the non-corporal conscious beings cooperate in leagues with others of their kind means that those in competition with them in their realm are colored as "good guys" and "bad guys" depending on their vantage point and vision for the direction of the beings in the material realm. Demons and Angels, each competitive side may be perceived as either one of these old designations depending on whose values are idealized. By these terms "God" for such creatures could be seen as the collective intellect personified. Whether a unified being is collectively seen or heard among them is unclear but it is clear that though there are competing ideologies for the best way to deal with the material real, and the creatures conveyed here, they do have a civility code among themselves. This is paralleled in the human reality in the Ten Commandment like boundaries or legal codes accepted by the majority of all people of all cultures everywhere though they derived their civility codes independent of one another.

Our Ethics Apart from Gods, Angels, Demons and any Other Kind of Entity

God, the idea of God among humans has to evolve to mean those qualities which we universally agree are good, noble, right, dignified, honorable, appropriate, inspirational, ideal. It is these qualities which we should find within ourselves. Humans need to stop separating their better qualities out and apart from themselves and then ascribing them outside themselves to the "other". All that is wise and good should not be seen exclusively in a being outside of ourselves. The fact that we can identify what is good, noble, honorable, dignified, ideal and generally right, yet relegate this outside ourselves means that we excuse and accept our evil lower qualities and actions as the qualities which better define us. This is wrong, this is a tragic mistake. When we personify good or evil as forces outside ourselves we wash personal responsibility away, this creates great crime. The human race is in danger of self-extinction so long as it keeps ascribing good and evil outside itself. People talk about being misled by the devil or evil and being rescued by the external good, Jesus, God, Allah or whatever other name of a deity they follow. Humanity has to grow up, take responsibility or its demise is certain and fully justified demise and this ends up being is a good thing. Humanity allows and excuses endless evil and abuse in our world with the expectation that god or a god/man or government will come and stop all the evil and turn the tide back toward the good. That's an awful big job, even for a god of limitless power and good. Further a race that is that callous does not deserve to be rescued. It is wrong to shirk off our personal responsibilities, we all should put evil and our more hateful sides in check and cull, cultivate and present, within ourselves, those more benevolent and cooperative qualities which help and demonstrate kindness of our own initiative. So to clarify, there both is and isn't a god. The collection of universally accepted good exists but it is not external to the field and ergo is found within all living things and should be yielded to.

It's not a god's job, or an angel's or a demon's job, to fix your messes. It's your job to limit the messes you make and to, conversely, actually show compassion, help, give kindness and cooperation aimed worldwide. Enlightenment looks like, sympathy, empathy, compassion, concern, kindness, help, cooperation, magnanimity, grace, governance of crude actions. Self-serving actions at the expense of everyone else's best interest, are ignoble and undignified. Be more, be intelligent, be educated, enlightened, be more than just a being of light whose intellect is retarded by the short lived material realm. As we know the universe is subject to the laws of thermodynamics and that entropy is the master of material reality. It is winding down, it will run down, stop, die, end, dissipate, experience heat death, and eventually blackout. The human race is intellectually impeded/retarded, it can't help it, and its consciousness is trapped in tangled, entangling/slowing/impeding networks of EMR knots. EMR thoughts and reality run up against the realm of mass. We are so much more, we should walk worthy of intellect and embrace it, the *"AND"* WORLD view, vs. the *"OR"* WORLD impeded ethos and view.

The "AND WORLD" view says our work should be for the mutual comfort and gain of all and not just for the most vicious. All the good things should be sought by all, for all, as a personal goal. We should say, I am rich, *AND,* you are rich. I am happy, AND, you are happy. I have most of what I need and want, AND, you too have most of what you need and want. We currently live in the, *OR ,* world where we have one person who is rich and another is homeless, beyond poor. Currently you are rich, OR, I am rich. You are comfortable, or I am comfortable. She owns a yacht, OR, I own a yacht. What is worse it is often the case that I am rich because you are poor, or you are poor to keep me rich. Currently the human race, even including our greatest scientist, currently are mere children in intellect, behavior, science, consciousness and ethics. It is time to be more and better. Our true selves, our conscious forms, around which DNA is molded, are configurations which have been around a very long time/eternally, but they are, not the same as they were long ago, they grow and gather to themselves and add to our whole being.

Finally Let's Square Up on One Main God and Jesus

So in final analysis, there is the Field and the Field produces all that is seen and unseen. The Field is omnipresent and generative, the Field is both conscious and unconscious, intelligent and unintelligent. The Field is tangible and intangible. So to answer the question, "Does God Exist?", in regards to the Field let's begin. The answer is both no and yes. The God of the Bible and of all the Holy books is a small, pale understanding and version compared to the definition of the Field itself. If you equate "God" with our understanding of the Field then, in these strict terms we can say, that this kind of "God" definitely exists.

The enigma of Jesus is just as hard to define. The better evidence suggests that Jesus was not a real person born as a flesh and blood person who lived out the scenes told to us in the New Testament. The stronger evidence suggests that this version of Jesus (or the Messiah/Christ) was a Roman coopted idea codified in stories composed by Roman Jewish scholars by the order of the Roman government. The story that they coopted however is taken from the most legitimate part of the Jewish Bible, or the fuller Tanakh. The largest and most reoccurring and driving theme of the Tanakh prophecies all point to the time of the coming of the God/man coming to set mankind straight. That time of prophecy culminated in the first century A.D. All the prophecies were pointing to that one time and at that time the Dead Sea Scrolls clarify that several such Messiah's arose at that time of revolt against Rome. Three prominent ones may have been amalgamated into the person later said to be "Jesus". One was known as Simon of Perea, another was Judah of Galilee and the third was Atronges of Emmaus. Some of these figures were more militant and some were more passive. Simon of Perea is the inspiration for the idea of Jesus dying and resurrecting. The point is, just as prophesied, the "spirit" of Christ, Jesus, the Messiah came at the requisite time and began to cause the requisite effect. The better qualities of man, ascribed to God or gods had come to a boil, a boil to the surface. The beginning of the betterment of man and woman had been initiated.

Because the war the Romans fought the Jews over was largely a Messianic war, the Romans coopted the Messiah belief to themselves to reshape it to be benevolent and pro-Roman. Their move was political and logical. This later afforded the Romans the opportunity to generate to benign religions, which they did. Christianity was the product of a coopted Jewish Messiah to the Roman Government and later church run by Romans separated and extracted completely from reorganized and redefined "Judaism". Judaism as is recognized today, is the antithesis of the historic religion of the Jewish and Israelite people which was one hundred percent messianic. The majority of the Jews were messianic in the first century. The war with Rome was a messianic motivated war and everything changed after that war. This being stated the ethics of peace were disproportionately emphasized in the New Testament as a social engineering project to rule the pro and the anti-messianic citizens of the Roman Empire. It worked better than war. Quite unintentionally the Romans ended up propagating the true spirit of the law, the messiah and Judeo-Christian ethics and values of charity and kindness, brotherhood etc. This is why that even people who know this history should value and promote Judeo-Christian values. Not because they are manipulated but because they agree with them because they are good. These ethics are the ethics of the Field itself.

The prophecies tell of two appearances of the Messiah/Christ/Jesus. This was also written into the New Testament composed by the pro-Roman Jewish Scholars as they codified the appearance of Titus, son of Roman Emperor Vespasian, in the role of the second coming Savior. But again, the second age of the Messiah has to tie back to a second awakening of the "spirit" of Christ in the human population once and for all.

It is time for the world to now cultivate the better versions of ourselves is also a product of the Field in its biggest evolutionary push of the Field and all of its agents in the end. Just because Jesus may not have been a real person, a flesh and blood man does not mean that "Jesus/The Messiah" is not REAL! Process this. The spirit of Jesus may have been a necessary culmination and component in the Field to better mankind's and the earths destiny. If the same point comes through as the conceptual Buda, Krishna or any other name for the inclusion and understanding of others of non-Christian cultures, then so be it. Grow in what you know.

Now you are in the know!

Bibliographical Derivation Sources

From Plato...

Plato: The Apology of Socrates

Plato: Crito

Plato: Meno

Plato: The Republic

Plato: The Timeous

Plato: Euthyphro

Plato: Gorgias

Plato: Laches

Plato: Lysis

Plato: Manexenus

Plato: Meno

Plato: Parmenides

Plato: Phaedo

Plato: Phaedrus

Plato: Protagoras

Plato: The Symposium

From Aristotle...

Aristotle: Physics

Aristotle: Metaphysics

Aristotle: Politics

Aristotle: De Anima (On the Soul)

Aristotle: Categories

Aristotle: Rhetoric

Aristotle: Nicomachean Ethics

Aristotle: Posterior Analytics

The Books Written by Tesla consulted...

My Inventions

Experiments with Alternating Currents of High Potential and Frequency

A Complete list of the U.S. Patents of Nichola Tesla

A Book about Tesla, Biography:

Margaret Cheney, "Man Out of Time".

The Book of Michael Faraday...

Experimental Researches in Electricity

The Book of Benjamin Franklin...

Experiments and Observation of Electricity made at Philadelphia in America

Books and Articles by Albert Einstein...

The Special and General Theory of Relativity

The Problem of Space, Ether, And the Field In Physics

Book of Stephen Hawking...

A Brief History of Time

Leonardo Da Vinci

Thoughts on Life and Art

Rene Descartes

First Philosophy

Secondary (Commentary) Sources...

The Greek Philosophers by

Dozen's of Youtube videos by Ken Wheeler on the nature of light, magnetism, electricity.

    Author of Uncovering the Missing Secrets of Magnetism.

Half a dozen Youtube videos of Eric P. Dollard, RCA Electrical Engineer and Genius on the nature

    Of electricity with the honor of being the only man to duplicate most of Nichola Tesla's

    Experiments with electricity and vacuum tubes. Dollard also is the man who has translated

The electrical equations of Sir J.J. Thompson the discoverer of the "electron" into standard

Modern notation.

Here is hoping that having read, studied and noted all the sources sited above I am at least fit enough to make sound observations regarding the title of this book based on these giants among men in thinking, science, philosophy and empirical observation, research and tangible inventions and productions.

www.ingramcontent.com/pod-product-compliance
Lightning Source LLC
Chambersburg PA
CBHW060838170526
45158CB00001B/185